INDUCTION, ACCEPTANCE, AND RATIONAL BELIEF

SYNTHESE LIBRARY

MONOGRAPHS ON EPISTEMOLOGY,

LOGIC, METHODOLOGY, PHILOSOPHY OF SCIENCE,

SOCIOLOGY OF SCIENCE AND OF KNOWLEDGE,

AND ON THE MATHEMATICAL METHODS OF

SOCIAL AND BEHAVIORAL SCIENCES

INDUCTION, ACCEPTANCE, AND RATIONAL BELIEF

Edited by

MARSHALL SWAIN
University of Pennsylvania

D. REIDEL PUBLISHING COMPANY / DORDRECHT-HOLLAND

Printed in The Netherlands by D. Reidel, Dordrecht

PREFACE

The papers collected in this volume were originally presented at a symposium held at the University of Pennsylvania in December of 1968. Each of the papers has been revised in light of the discussions that took place during this symposium. None of the papers has appeared in print previously. The extensive bibliography that appears at the end of the volume was originally distributed during the symposium and was revised on the basis of many helpful suggestions made by those who participated.

The symposium was made possible by a grant from The National Science Foundation and funds contributed by the Philosophy Department of the University of Pennsylvania. On behalf of the contributors to this volume, I would like to express my thanks to these organizations for their generous support.

In addition, I would like to express my gratitude to the members of the Graduate Philosophy Students Organization at the University of Pennsylvania for the considerable assistance they gave me during the symposium. My thanks, also, to Judith Sofranko and Lynn Luckett for their very responsible efforts in the preparation of the manuscript. Finally, I would like to thank Professor James Cornman for his invaluable advice and encouragement.

M. S.

TABLE OF CONTENTS

EDITOR'S INTRODUCTION

The papers collected in this book are all devoted to the problem of the role that induction plays in theories of knowledge and rational belief. Nearly all of the papers are programmatic, and yet they contain many specific and original suggestions toward solutions of isolated problems in the area.

The volume opens with an essay by Frederic Schick in which he delineates and compares three apparently competing theories about the conditions under which we are obligated or entitled to believe propositions. Schick calls these theories *subjectivism, empiricism*, and *pragmatism*; he refers to the writings of H. E. Kyburg, Jr. as one source of the subjectivist theory (not to be confused with the subjective theory of probability!), to the work of J. Hintikka as an example of empiricism, and to the work of I. Levi as an example of pragmatism. Schick argues that each of these theories is operative only in the context of a certain type of issue; empiricism and pragmatism provide principles relevant to fundamentally different (indeed, incompatible) types of issues, and therefore the results of these theories are not comparable. Consequently, the theories are not incompatible. Moreover, Schick argues, subjectivism is most plausibly interpreted in such a way that, depending on the issue, it will yield the same results as the other two theories. He suggests that a truly comprehensive theory of rational belief will incorporate all three of these theories.

Schick also distinguishes the principles that individuate each of these theories from certain other principles which the theories might have in common; the latter are called 'overarching principles'. For example, the principle that the set of propositions one believes ought (rationally) to be deductively closed might be an overarching principle. The subject of overarching principles is the topic of my paper, which follows Schick's essay. In this paper, I construct a general theory of rational belief, which is intended to provide, among other things, principles governing the admissibility of propositions into sets of rationally believed propositions. A distinction is made between propositions one is obligated to believe and propositions one is permitted to believe, from the point of view of rationality. I argue that the principles governing obligatory belief are quite different from those that govern permissible belief; this difference is illustrated in a solution given to the lottery paradox.The theory that I construct is intended to be 'overarching' in that it hopefully can serve as a framework within which rules of rational acceptance can be constructed. The theory is presented as an extension of Hintikka's logic of knowledge and belief, although many of the basic concepts of his system are considerably revised.

The problem of overarching principles is also the subject of the essay by Henry E. Kyburg, Jr. Kyburg discusses several general principles that have been adopted, either implicitly or explicitly, by a number of writers on acceptance. He distinguishes 'weak' and 'strong' versions of principles governing the consistency as well as the deductive closure of sets of rationally accepted statements; in addition, he considers a principle of conjunction, which stipulates that if statements s_1 and s_2 are rationally accepted, then their conjunction ought to be accepted. After tracing a number of connections that hold among these various principles, Kyburg considers acceptance rules proposed by Hempel, Hintikka and Hilpinen, Levi, and Lehrer, all of which satisfy the conjunction principle and some or all of the other principles as well. Kyburg argues that each of these systems has flaws, and that many of the flaws are most plausibly attributable to the fact that they satisfy the conjunction principle. He points out that his own system of rational belief, developed in earlier writings, abandons the conjunction principle and is consequently able to avoid many of the problems that the other systems face. Although his primary concern is to cast doubt on the conjunction principle, Kyburg also presents us with lucid expositions of the above theories and discusses many of their interesting aspects.

Overarching principles of the type discussed above have an obvious relevance to the problem of constructing an adequate acceptance rule. Once these principles have been agreed upon, they can be used to formulate a partial criterion of adequacy for acceptance rules; any adequate acceptance rule must at least satisfy the demands of the overarching principles. In his contribution to this collection, Gilbert Harman introduces another kind of consideration that is relevant to the adequacy of acceptance rules; namely, the connection between inferences warranted by such rules and the analysis of knowledge. Harman suggests that one partial test of adequacy for a proposed acceptance rule is whether inferences in accordance with that rule could provide one with knowledge. He discusses the causal theory of knowing developed by Alvin Goldman, and points out that this theory suggests a theory of inductive inference; namely, that inductive inference is inference to the best of competing causal statements. That theory has defects, however, and Harman proposes that we replace it with the theory that inductive inference is inference to the best of competing *explanatory* statements, a view that he has developed in previous writings. In the present paper, Harman expands and clarifies this interesting doctrine.

The relation between explanation and justified belief is also the subject of the highly original essay by Keith Lehrer. Lehrer suggests that a belief is justified either because of what it explains or because of what explains it; he dubs this theory of justification *explanationism.* Warranted inductive inference is viewed by Lehrer as one source of justified belief; if a hypothesis h is inducible from one's background knowledge, then h is explained by that knowledge, and given the theory of explanationism, it follows that h is justified by that background knowledge. Lehrer provides us with a number of overarching principles governing the explanation of a hypothesis or set of hypotheses relative to given background knowledge, and constructs an inductive acceptance rule such that inductive inferences warranted by that rule satisfy the demands of these overarching principles. Lehrer then argues that hypotheses inducible in accordance with this rule are also explained by the background knowledge from which they are induced. Consequently, he argues, his rule satisfies the demands of explanationism; that is, a hypothesis is justified by inductive inference in accordance with his rule only if it is explained by the background knowledge. In the course of his discussion, Lehrer sheds new light on many interesting problems; for example, he discusses the problem of self-justified (but fallible) beliefs within the context of explanationism.

The papers discussed thus far have been directed primarily at problems surrounding the concept of rational acceptance, or rational belief. It is often supposed that a theory which assumes that there is such a thing as acceptance is fundamentally opposed to theories which instead speak of (rational) degrees of belief, or degrees of credence. In his important essay,

Isaac Levi argues that this is not necessarily so. He points out that a theory of rational probability judgement will consist of conditions that determine rational degrees of credence at a given time as well as rational revision of degrees of credence over time. He rejects the Carnapian image of such conditions as a set of norms for an ideally rational man on the grounds that such a view has untoward consequences. He suggests, instead, that the conditions for rational probability judgement be viewed as necessary conditions for a state of 'equilibrium', analogously to a comparative static theory of consumer demand. In order to avoid subjectivism, Levi argues that the conditions for rational probability judgement must ground such judgements empirically, and he suggests a way in which this can be done. The resulting theory, he points out, presupposes that it is legitimate for one to accept some statements as evidence, contrary to a proposal made by Richard Jeffrey that rational credence judgements can be empirically grounded even though no statements are accepted as evidence. Levi concludes that an adequate empiricistic theory of rational probability judgement must allow for the acceptance of statements as evidence; and, in turn, an adequate theory of acceptance will presuppose a theory of problem solving, which utilizes rational probability judgement. Thus, rather than being incompatible, the theory of rational credence judgement and the theory of acceptance are mutually dependent.

Isaac Levi is to the theory of acceptance as Richard Jeffrey is to the theory of partial belief. In his contribution to this collection, Jeffrey offers a defense of his approach to the problem of rational belief; he is concerned especially to answer some standard criticisms of his approach (some of which have been raised by Levi). The Bayesian theory espoused by Jeffrey is, he points out, a pragmatic theory intended to provide a framework for rational deliberation which issues in action. The theory uses the term 'degree of belief' as a technical term which refers to an attribute of ones' attitude toward risks; the ordinary sense of belief, which seems to be used by those who speak of acceptance, is only vestigially present in this concept. One of the conditions that a set of partial beliefs must satisfy if it is to be rational is conformity to the laws of the elementary probability calculus. The theory is often criticised on the grounds that if this is sufficient for rationality, then any coherent set of partial beliefs will do in a given situation. But, Jeffrey argues, this is a misunderstanding;

he suggests that coherence is at best a necessary condition for rationality, that there are surely other necessary conditions, and he provides us with some specific examples in which considerations (such as empirical facts) that go beyond mere coherence are relevant to the rationality of a set of partial beliefs. In addition, Jeffrey offers a defense of his kinematical theory regarding the rational revision of one's belief function against some criticisms raised by Levi, and sheds light on a number of other aspects of his theory. Jeffrey concludes his defense, as did Levi, in a spirit of reconciliation; perhaps theories of acceptance and of partial belief are not in fact mutually exclusive but are rather complementary parts of a comprehensive theory of rational belief and action.

Although these papers deal with a wide range of problems in the area of rational belief, there are naturally many problems that do not come under discussion. The bibliography prepared by Ralph Slaght, which appears at the end of the volume, is designed to guide the researcher toward recent literature on the subject of this book and on a number of related subjects.

FREDERIC SCHICK

THREE LOGICS OF BELIEF

From the point of view of the practical logician, we start in the middle of things. We consider what it is we ought to believe in the context of some issue that concerns us. In this paper I shall distinguish several sorts of such issues and shall argue that the basic current logics of belief are each serviceable with regard to one of these sorts of issues only. My conclusion will be that these logics are not incompatible with one another, as is usually supposed, but rather mutually supplementary, that a comprehensive analysis of rational belief must incorporate them all.

In Sections I–III I shall formulate three logics I shall label *subjectivism, empiricism* and *pragmatism*. The logic I shall describe as subjectivist is a pruned and reshaped version of Kyburg's analysis. I shall take Hintikka's position as my point of departure for empiricism and focus upon Levi's analysis in connection with pragmatism. In Section IV I shall consider how these logics relate to one another. I shall wind up with a few comments on some very general principles most authors adopt.

I. SUBJECTIVISM

The subjectivist analysis holds that we may believe a proposition if and only if it fits in with those we already believe, and that we *must* believe it if and only if avoiding the belief would make for some imbalance in the set of all our beliefs. Suppose that q is the least well confirmed proposition, relative to my information, that I believe, or any one of these propositions if there are several such. I suggest that the set of all my beliefs is *balanced* if and only if I display no bias against any proposition, if and only if I believe all those propositions under consideration that are at least as well confirmed, relative to my information, as q is. (Analogously of course for *your* beliefs, your cousin's, your brother-in-law's, etc.) A subjectivist logic of belief can be phrased as follows:

(S) A person believes p if and only if $c(p, i) \geqslant \omega$, where p is some proposition the person considers believing, i is his information, and ω is the degree of confirmation of the least well confirmed proposition he believes.

This makes for a logic in the traditional, modest sense. Principle S – as also each of the others to be presented – sets forth a necessary condition of rational belief in the contexts in which it applies. It does not identify a sufficient condition. It provides a criterion of what a person is rationally committed to believing in the appropriate contexts, but not an exhaustive criterion.

The limits of the scope of applicability of S call for some attention. S imposes commitments with regard to live issues only. It does not require anyone to take a stand in connection with issues he is not raising with himself. More specifically, the applicability of S is restricted to situations in which a person considers believing some single, isolated proposition, to cases in which he is asking *ought I to believe this proposition or not.* Whether or not to believe a proposition is the characteristic form of a certain familiar sort of issues. I shall refer to issues of this sort as *questions.*[1]

The alternative to believing a proposition is not *dis*believing it (believing it to be false) but rather *not* believing it (either believing it to be false or suspending judgment). It may often be difficult to determine whether an issue is one of believing p or *dis*believing it or one of believing p or *not* believing it. But the distinction can be made. We are involved in a situation of the latter sort – that is, we face a *question* – when we suppose that disbelieving p would be of no significance to us, would neither help us to explain nor to clarify nor to confirm nor to discredit anything that currently interests us. A case in point is Miller's discussion [13] of the hypothesis that 7 ± 2 is the limit of the number of discriminations we are capable of making within any unidimensional category. He asks whether or not he ought to believe this hypothesis, not whether he ought to believe it or to disbelieve it. For though the hypothesis itself is startling and would explain a considerable amount of data, its contradictory is too broad to be of any service. (If 7 ± 2 is not the limit, is it 8 ± 2, or 37 ± 6, or is there no limit at all?) Miller accepts the hypothesis, but had he not accepted it, he would probably not have bothered to believe it to be false. He would have suspended judgment on the matter.

The gist of S is that our commitments depend on our involvements. If we believe q, and p is at least as well confirmed, relative to our information, as q is, then if p comes up for consideration, we are committed to believing it. If we do not believe any proposition whose degree of confirmation is as low as that of p, we need *not* believe p. Our epistemic involvements are always subject to reassessment, and so we need not, on this theory, believe any proposition we would prefer not to believe. We need only make sure that we do not believe any propositions whose degree of confirmation is as low as that of one we wish to set aside. We need not in fact believe anything whatever. From the point of view here taken, a thoroughgoing skeptic cannot be faulted. He is not remiss in his beliefs. If a person has no involvements, he has no commitments.

S precludes our believing any propositions whose degree of confirmation is less than ω. But this is a vacuous restriction, since ω is the degree of confirmation of the least well confirmed proposition we believe, and nothing we believe can consequently fail to pass muster.[2] Prospectively there is some dissuasion here, but retrospectively anything goes. Only non-beliefs can establish an imbalance. This has striking consequences, for if a person is sufficiently easygoing in his beliefs – if, for him, ω is 0.5 or less – he may be committed to believing incompatible propositions. Indeed S does not even rule out his believing propositions incompatible with his information, and if he does believe some such proposition, he is committed to believing every other proposition that comes into question for him. An indiscriminate credulity cannot be faulted any more than total skepticism.

We can provide against these new excesses by qualifying S. We need only stipulate that ω be greater than 0.5. I prefer, however, to keep S as it stands. This principle identifies a necessary condition of rationality in issues of a certain sort. Many authors would argue that there are also criteria of rationality that are invariant with regard to the sorts of issues involved. One of these – I shall call it the Principle of Coherence – is that no proposition be believed that is not more likely than its contradictory, and this rules out believing incompatible propositions and propositions incompatible with our information. (The Principle of Coherence precludes our believing any propositions whose degree of confirmation is not greater than 0.5, and thus indeed restricts us to values of ω greater than 0.5.) The proposed overarching criteria are not distinctive of any specific logic of belief. (This is what makes them overarching.) There is no reason to build any of them into the principles we are developing here. But since the Principle of Coherence is at any rate likely to be acknowledged, we need not be disturbed by any unorthodox implications of our special principles considered independently. (More on coherence in Section V.)

S is phrased in terms of a quantitative concept of confirmation. This is not essential to the case. The principle can be rephrased so as to refer to comparisons of confirmation only:

(*S**) A person believes *p* if and only if there is some proposition *q* he believes such that *p* is at least as well confirmed, relative to *i*, as *q* is,

where *p* is some proposition the person considers believing and *i* is his information. Putting the matter in these terms points up the narrow scope of our methodological assumptions. *S** does not presuppose the quantitative measurability of confirmation. It does not even presuppose that propositions are always comparable with regard to confirmation – that, of any two propositions, the first must be either better or equally or less well confirmed, relative to the information, than the second. If two propositions are not in fact comparable, then *S** simply cannot commit anyone to believing either on the basis of his belief in the other.

A word on information. It is useful to distinguish two concepts here, that of the information a person *has* and the more inclusive notion of the information *available* to him (the information *to be had*). The latter is vague. How difficult must it be for a person to acquire an item of information before we will say that it is not available to him? There are, at any rate, degrees of availability, and this complicates matters. It promises to be more fruitful to consider how the rationality of a person's beliefs is related to the information he *has*, and this is therefore how I shall understand all references to information in this paper.

About this concept itself I have little to say. The information a person has is a body of propositions. All the items of a person's information are moreover believed by him. (This is the crux of the distinction between the two concepts of information.) The converse is false – it is not the case that all the propositions a person believes are items of his information. Only those propositions he is entitled to offer in support of the propositions he believes are informational for him, or rather only those propositions which, if he believed them, he would be entitled to consider supportive. I remark on this only to forestall misunderstandings. It does not take us far. I wish I could do better. But the reader will have to shift for himself here.

II. EMPIRICISM

The point of departure for empiricism is the unattainability of certainty, or rather the illegitimacy of any pretensions to certainty with regard to the world at large. We raise issues for ourselves which are bound to remain conjectural. The empiricist argues that since we must at any rate conjecture, we might as well do so in a manner that invites at least our own confidence. As he sees it, this requires that we take probability as our guide, and let the observations we have made of the world be the basis for the assessment of the probabilities involved. He holds that a rational person commits himself only in the face of high probabilities backed by the evidence of his observations. In a prolix but useful formulation, the empiricist holds that a rational person believes a proposition if and only if it has a sufficiently high degree of confirmation on the basis of an adequate number of serviceable observations. An empiricist logic might tentatively be phrased as follows:

(E_H) A person believes p if and only if $c(p, i) \geqslant 1 - \varepsilon$ and $n \geqslant n_0$.

Here p is any proposition, i is the person's information, n is the number of serviceable observations he has made, and n_0 and $1-\varepsilon$ are two measures of his epistemic caution – the former identifying what he considers to be an adequate number of observations, the latter what he considers a satisfactorily high degree of confirmation.

Principle E_H is adapted from a principle proposed by Hintikka and Hilpinen (whence the subscript).[3] These authors go on to suggest a way of tightening the analysis, a way of reducing the number of independent indexes of caution to one. Their discussion here is based on Hintikka's previous revision of Carnap's theory of confirmation ([4], [5]). On Carnap's theory, the *a posteriori* degree of confirmation of every universally general proposition remains vanishingly small until the number of observations made approximates the number of individuals in the universe. Hintikka provides for intuitively more appropriate confirmation-values. In particular, it turns out, on his analysis, that as the number n of observations of exemplifications of Q-predicates increases, so also (fairly rapidly) does the degree of confirmation of the generalization asserting that every individual in the universe is characterized by one of the Q-predicates exemplified in the sample, the degree of confirmation of this proposition approaching one as a limit. (A Q-predicate is a compound predicate formed by taking either the atomic predicate or its complement from every basic dichotomy in the language, and conjoining the lot.) Since the *a priori* confirmation of this generalization is close to zero, there must be a point in the sequence of observations at which its degree of confirmation first both exceeds ε and equals or exceeds $1-\varepsilon$. This is the point at which our information first assigns to the negation of this generalization a degree of confirmation precluding its belief while affording the generalization itself a degree of confirmation *not* precluding belief. It seems proper to hold that when this point is reached, when our information first confirms to this high a degree that we have seen all that there is to see, we have put together a body of information that is adequate for inferential purposes. We have, that is, made an adequate number of observations. Hintikka and Hilpinen accordingly propose to let n_0 be the lowest number such that, if $n \geqslant n_0$, the degree of confirmation of the generalization asserting that every individual in the universe is characterized by one of the Q-predicates exemplified in the sample is both greater than ε and equal to or greater than $1-\varepsilon$.[4]

This analysis is suggestive, but I do not think it will stand. The difficulty has to do with the concept of serviceability. A set of serviceable observations need not always be a set of Q-observations. (Hintikka and Hilpinen do not explicitly say that it must, but they consider no other sorts of observations.) And a case can be made for holding that serviceability is best thought of as a characteristic of *sets* of observations, and that serviceability *simpliciter* is not enough in any case.

The serviceability of an observation depends on the problem at issue and on the information we have. If we are asking whether beavers mate in the spring or the fall, observations of the behavior of elks are not serviceable. Where we are concerned with the behavior of elks, our observations of beavers are not serviceable. Our observations of beavers are not serviceable even in the former case if, in the context of our information, they are compatible with both of our hypotheses, for we are looking for grounds on which to accept one rather than the other, or neither. If, relative to our information, an observation is incompatible with at least one of our hypotheses, then it *is* serviceable. In sum: an observation is serviceable with regard to a problem in the context of a body of information if and only if, when conjoined with that information, it falsifies some of the hypotheses involved.[5] We shall make use of a closely related characteristic of *sets* of observations. First some new technical terms. Let me call a set of propostions a set of *basic alternatives* for a person if and only if each is acceptable to him as a means of settling an issue that currently concerns him and none is a deductive consequence of any of the others and there are no other propositions of this sort he finds acceptable for his purpose. (Believing a basic alternative is a maximally affirmative option, not necessarily the option a person thinks or should think the best available.) Consider now a set of basic alternatives whose members are pairwise exclusive relative to the information this person has but not collectively exhaustive relative to that information. I shall call a set of basic alternatives of this sort a *problem-set*. Every *problem* – this too is a technical term here – is characterized by a single problem-set. A set of observations may be said to be serviceable with regard to a problem in the context of a body of information if and only if it contains, for at least one proposition in the problem-set characterizing that problem, at least one observation which, when conjoined with the information, falsifies that proposition. A set of observations will, finally, be called *strictly* serviceable with regard to a problem if and only if it contains, *for every proposition but one* in the problem-set characterizing that problem, at least one proposition which, when conjoined with the information, falsifies that proposition.

Note that if a set of observations is strictly serviceable with regard to a problem, it is also serviceable *simpliciter* with regard to it. More important, if we call one problem a *sub-problem* of another when each of the basic alternatives of the first are alternations of some of those of the second – which would, for instance, make $\{p\text{-or-}q,\ r\text{-or-}s\}$ a sub-problem of $\{p, q, r, s,\}$ – then a set of observations that is strictly serviceable with regard to one problem is serviceable *simpliciter* with regard to any other of which the first problem is a sub-problem.

A set of basic alternatives is nonexhaustive where the complementary alternative – the one that would exhaust the domain of possibilities – is of no interest to the person raising the issue. A simple example is provided by Köhler's classic experiments with chickens (see [9]). Köhler wanted to determine whether chickens taught to single out the darker of two grays discriminated specific colors or relative darknesses of color. These were his basic alternatives. They were not conjointly exhaustive: it may be that some chickens discriminate specific colors and others relative darknesses, or that all chickens discriminate both, or that none of them discriminates either, or that some still other circumstance obtains. But the alternation of all these possibilities did not interest Köhler. Adopting this compound proposition would have served none of his theoretical purposes. (At least at first; see [9], pp. 165–7.) Thus the issue he raised was (technically) a problem.

We must still consider how we might determine whether an *adequate number* of observations has been made. A simple course is available. Let me suggest that an adequate number of observations has been made with regard to a problem if and only if the set of these observations is serviceable with regard to that problem. On this analysis, the requirement of adequacy turns out to be redundant.

The logic of empiricism may now be formulated as follows:

(*E*) A person believes p if and only if p is the strongest proposition of which it is true both that (a) $c(p, i) \geqslant 1 - \varepsilon$ and that (b) the set of the observations made is strictly serviceable, in the context of i, with regard to any problem characterized by $\{p, q, \ldots, z\}$,

where p is an alternation (possibly a one-fold alternation) of some members of the problem-set actually at issue, $q, \ldots, z$ are the remaining members of that problem-set, and $1-\varepsilon$ and i are as before. (It may be that none of our basic alternatives satisfies both of the conditions cited in *E*, but that some two-fold or three-fold or four-fold alternations of them do. *E* commits us to believing the strongest of these eligible propositions.)

Principle *E* can be regarded as a new interpretation of E_H, n being identified as the number of observations (of whatever sort) made, and n_0 as the number of observations made at the first point in the sequence of these observations at which the set of them all is strictly serviceable, in the context of i, with regard to any problem characterized by $\{p, q, \ldots, z\}$. If, however, we think of E_H as Hintikka and Hilpinen themselves do – that is, in terms of their own interpretation of n and of n_0 – the two principles are distinct. They are in fact not comparable. Either may commit (or entitle) us to believe a proposition without the other committing (or entitling) us to believe it. In what follows, I shall take E and E_H to be two distinct principles.

One feature of our analysis calls for special comment. The propositions constituting a problem-set cannot (by definition) be collectively exhaustive relative to the information of the person involved. Why this unorthodox restriction? Why could we not assign E a more inclusive applicability? Suppose that the restriction did not hold, and consider a case in which the person's information did imply that one of his basic alternatives was true. If his observations, in conjunction with his information, falsified some of these alternatives, the alternation of the ones that remained would be a deductive consequence of his information subsequent to the observations. E would thus willy-nilly limit a person to deductive inference here. There would be nothing wrong with this extremity of caution if it were self-imposed. Indeed, by allowing ε to be zero, we provide for adopting it as a general way of life. But there should be no constraint in the matter. A person should be free to exercise a lesser degree of caution in every problem that concerns him. Stipulating that the elements of a problem-set not be collectively exhaustive mandates this freedom by limiting the scope of the concept of a problem. (This is shifty, but innocuous. We shall return in the next section to issues of the sort ruled out here, and shall see that, in connection with an appropriate logic, we may adopt any degree of caution with regard to these issues too.)

III. PRAGMATISM

The pragmatist's thesis is that our actions are rational to the extent that they contribute to the furtherance of our ends. He considers belief to be a species of action, and thus holds that we must assess our beliefs too in these terms. He may go on to contend that the epistemic objectives of an unprejudiced inquirer are distinctive in some way. In that case he will maintain that rationality requires us to focus on these special objectives. Isaac Levi has developed by far the most elaborate analysis in this vein. (See [11].) Let me outline his position.

Levi holds that a rational person is motivated in inquiry by two objectives. He wants both to discover truth and to relieve doubt. The first of these needs no analysis in this context. The second is more obscure, and Levi provides a useful discussion of the matter. He is concerned not with the relief from doubt *provided*, but with the relief from doubt *offered* by belief, and indeed not with relief from doubt proper, but with relief from the doubt our information entitles us to maintain. This is an affair of logic rather than of psychology. The relief from doubt offered must nonetheless be offered to someone. How much relief is offered that person depends on the inquiry in which he is engaged.

Consider again a set of basic alternatives, the propositions this time being both pairwise exclusive and collectively exhaustive relative to the information of the person involved. Symmetry suggests that we refer to such sets as *inquiry-sets*. Suppose that m of the members of some inquiry-set are compatible with the person's information. Determine the number of these m propositions to the alternation of which some proposition p is equivalent, given the information, and divide this number by m. One minus the resultant quotient is the *content* of p, relative to that information. This is Levi's index of the relief from doubt that believing p offers a person with this information in the context of the inquiry at issue. The relief from doubt offered by a belief is thus measured by the specificity with which the adoption of that belief would, under the circumstances, resolve the inquiry at issue.

The difference between the issues here labeled *inquiries* and those we have called *problems* should be kept in mind. The elements of a problem-set are not collectively exhaustive, given the information involved, while the elements of an inquiry-set *are* exhaustive. As an example of an inquiry, consider the early work of the Würzburg School on the subject of imageless thought (see Humphrey [8], Chapter 2). The Würzburgers' issue was: are there any imageless thoughts or are all thoughts accompanied by images? The basic alternatives here are mutually exclusive and collectively exhaustive.

Having discussed the nature of our objectives in inquiry, Levi turns to an analysis of the amalgamation of these objectives. What he wants is a measure of the utility (all things considered) of believing the propositions that concern us. He first proposes some theses concerning the relations of the utilities of believing various sorts of propositions. Combining these theses, he arrives at $1-\eta\,\mathrm{cont}(\text{not-}p, i)$ as the measure of the utility of believing p when p is true and i is the information, and at $-\eta\,\mathrm{cont}(\text{not-}p, i)$ as the measure of the utility of believing p when p is false and i the information. Here $\mathrm{cont}(p, i)$ is Levi's measure of the content of p

relative to i, and η (Levi uses 'q') is an index of the epistemic boldness of the person involved. (Levi sets $0<\eta\leqslant 1$. For the sake of generality, I shall set $0\leqslant\eta\leqslant 1$. Nothing of substance is affected.)

The expected utility of believing p, relative to information i, can now be determined (in accordance with the standard definition) by adding the product of the degree of confirmation of p and the utility of believing p when it is true, to the product of the degree of confirmation of not-p and the utility of believing p when it is false – all of course relative to i. Some simple calculations yield $c(p, i)-\eta\,\text{cont}(\text{not-}p, i)$ as Levi's formula for the expected utility of believing p, relative to i. This provides Levi with the utility-measure he is seeking. (More precisely: pending the specification of a value of η, it identifies a *family* of such utility-measures.) Levi now endorses the following principle:

(P_L) (1) A person believes p if either (a) the expected utility of believing p, relative to his information, is greater than that of believing any other proposition q, or (b) p is the alternation of those propositions that have the maximal expected epistemic utility, if there are several such,
(2) if he believes p, he believes all the deductive consequences of the conjunction of p with his information, and
(3) he believes nothing else.

Both p and q are here either members of one of this person's inquiry-sets or alternations of some members of that set.

Principle P_L is not distinctively Levi's. The policy of maximizing the expected epistemic utility is, in some form or other, the one generally proposed by authors developing a pragmatist logic of belief.[6] What is distinctive here is Levi's analysis of epistemic utility, which provides for specific directives in accordance with P_L, and particularly his introduction of an index of boldness. (Roughly speaking, the larger our η, the greater the content, relative to our information, of the proposition assigned the maximal expected epistemic utility, and hence the greater the content of the strongest proposition P_L entitles us to believe.) But this analysis is independent of P_L itself. There are alternate conceptions of epistemic utility, and P_L accommodates these too.

Let me propose a somewhat weaker principle, and take the occasion to rephrase the maximizing policy itself. Consider the alternation of those of a person's basic alternatives that are compatible with his information. Since the basic alternatives are collectively exhaustive relative to the information, to believe this alternation and no sub-alternation is to suspend judgment on the issue. This suspension of judgment may in turn be regarded as the maintenance of the epistemic *status quo*. One ought clearly to try for the greatest possible improvement on the *status quo*. On Levi's analysis, the expected utility of remaining where we are – as also that of believing a tautology – is always $1-\eta$. (The bolder we are, the lower the expected utility of suspending judgment.) The following principle recommends itself:

(P) A person believes p if and only if either (a) the excess over $1-\eta$ of the expected utility of believing p, relative to the person's information, is greater than the corresponding excess for every other proposition q, or (b) p is the alternation of those propositions that are characterized by the maximal such excess, if there are several of them.

Both p and q are again either members of one of this person's inquiry-sets or alternations of some members of that set.

The if-clause embedded in P is equivalent to the first clause of P_L, given Levi's analysis of epistemic utility. The two principles differ with regard to deducibility. There is nothing in P corresponding to the second clause of P_L. In consequence, P is less demanding (and more restrictive) than P_L. If P commits (or entitles) us to believe a proposition, P_L commits (or entitles) us to believe it too, but the converse does not hold. Let me here say only that I count this in favor of P. (I discuss deducibility requirements in Section V.)

The rationale offered for P suggests also a simpler principle. Consider again the alternation of those of a person's basic alternatives that are compatible with his information. As noted above, believing this alternation and no sub-alternation maintains the *status quo*. It might be argued that whatever has a higher utility is worth having, and that maximizing is therefore not enough. If the expected epistemic utility of a proposition is greater than that of the alternation described, we ought to believe that proposition. And since a rational person is bound to acknowledge this alternation itself, we must believe those propositions whose expected epistemic utility is just equal to that of this alternation. A pragmatist might propose the following principle:

(P') A person believes p if and only if the expected utility of believing p, relative to his information, is equal to or greater than $1-\eta$,

where p and $1-\eta$ are as before.

P' is obviously more demanding than P. Indeed it may seem excessively demanding, for if η is sufficiently large, P' might commit us to believing incompatible propositions in the context of some single inquiry. (Believing p and believing q may both, independently, have an expected utility greater than that of suspending judgment where the expected utility of believing p and believing q conjointly is zero.) We encountered an analogous situation in our analysis of S. At that occasion, we noted that the adoption of the Principle of Coherence prevents us from abusing the freedom that S grants. So also here. The Principle of Coherence restricts us to values of η less than 0.5 (see below), and this cuts off all conflicting commitments. There is, however, still something wrong here. Although the conjunction of P' with the Principle of Coherence is only slightly more demanding (and slightly less restrictive) than the conjunction of P with that principle, P' does require too much. P is, I believe, the sounder principle. (More on this matter in Section V.)

The remarks leading up to the introduction of P hinge upon the fact that, on Levi's analysis, $1-\eta$ is always the expected utility of suspending judgment. This is not the case on any of the other current analyses of epistemic utility. But the adoption of P is nonetheless not tied to the adoption of Levi's analysis. Hilpinen [3] accepts $1-\mathrm{c}(p)$ as the measure of the content of p, and obtains $\mathrm{c}(p, i)-\eta\,\mathrm{c}(p)$ as the expected utility of believing p, relative to information i. On his analysis, $1-\eta$ is the expected utility of believing a tautology. It may but need not also be the expected utility of suspending judgment. Where the two are distinct, the expected utility of suspending judgment is the greater. The adoption of P along with Hilpinen's analysis of epistemic utility instead of Levi's thus involves setting the expected utility of suspending judgment either on or above the threshold of commitment in preference to identifying it as that threshold itself. The suspension of judgment remains the option of last resort. The outlines of the logic are unchanged. Hilpinen's analysis is, however, only obliquely responsive to variations in the size of the inquiry-set, and its specific directives are consequently likely to be less convincing than those that Levi would offer. We shall therefore continue to work with Levi's analysis.

Must we restrict the application of *P* to inquiries? Could not the logic of inquiries be expanded so as also to do the work now assigned to the logic of problems? Suppose that some issue we have raised is (technically speaking) a problem – that our basic alternatives are pairwise exclusive but not collectively exhaustive relative to our information. This means that the complementary proposition is of no interest to us, that believing this proposition would serve no epistemic purpose. Since the basic alternatives are the only propositions we accord any independent significance, believing the alternation of all those that are compatible with our information and no sub-alternation is merely to hold that our problem is worth considering. We ought, in the proper circumstance, to be entitled to take this stand. But we are never entitled to do so by *P* given Levi's analysis of epistemic utility. On that analysis, the expected utility of believing the alternation of all the basic alternatives here that are compatible with our information is less than $1-\eta$, and believing this alternation is therefore always ruled out. (We might of course revise *P*, taking the expected utility of believing the alternation of all those of our basic alternatives that are compatible with our information, whatever their collective scope, as a variable threshold of commitment, but in that case we would *always* be committed to believing some alternation of these alternatives, and this would also be unacceptable.) I conclude that either Levi's analysis of epistemic utility or else indeed the maximizing policy itself, is not suited for use where the basic alternatives are nonexhaustive, and that we must consequently restrict the application of *P* to situations in which our basic alternatives *are* exhaustive – that is, to situations in which our issue is an *inquiry*.

IV. THE COMPATIBILITY OF THESE THREE LOGICS

P commits (and entitles) us to believe *p* only if $c(p, i)-\eta$ cont (not-p, i) $\geqslant 1-\eta$. Now if $c(p, i)-\eta$ cont(not-p, i) $\geqslant 1-\eta$, then $c(p, i) \geqslant 1-\eta$, for η cont(not-p, i) must be either positive or zero. Thus *P* commits (and entitles) us to believe *p* only if $c(p, i) \geqslant 1-\eta$. We may recall that *E* commits (and entitles) us to believe *p* only if $c(p, i) \geqslant 1-\varepsilon$. The parallelism here is striking, and all the more so when we note that we identified $1-\varepsilon$

as an index of caution. Both ε and η are indexes of boldness, and the extent of our boldness, as measured by these indexes, is assigned a similar significance in the two logics in which they figure. This raises the question whether they are in fact separate indexes, whether, that is, we need them both.

How might the limits of our boldness be recognized? Consider η. Since boldness is reflected in the risks one is willing to take, our personal η might be thought to be identifiable by the distance down the scale of utility we are willing to go in the adoption of beliefs. But this course is closed, for η itself figures in the determination of expected utilities. A natural alternative is to consider how far down we are willing to go with regard to degrees of confirmation. In adopting *P*, we bind ourselves to believing no propositions whose degree of confirmation is less than $1-\eta$. The greater our appetite for belief, the lower we must go. Let us therefore note the point in the downward progression at which we would call a halt. The lowest degree of confirmation to which we are ready to descend – or rather, the lowest degree of confirmation to which, in some context, we are in fact descending – marks the limit of the risks we are willing to tolerate in our beliefs. To accept this analysis is to identify one minus the degree of confirmation of the least well confirmed proposition we believe as the measure of our boldness here. More directly: it is to identify that degree of confirmation itself as the measure of our caution.

Analogously for ε. One minus the degree of confirmation of the least well confirmed proposition we believe can be chosen as the value of ε as well as of η. In our discussion of the subjectivist position, we marked this critical degree of confirmation ω. If we accept the interpretations of η and of ε suggested, we now have $\eta=\varepsilon=1-\omega$, or $1-\eta=1-\varepsilon=\omega$. Not only do we not need both η and ε. It turns out we need neither.

This allows for some economizing. We can, for instance, formulate a comparative version of *E* on the model of S^*:

(E^*) A person believes *p* if and only if *p* is the strongest proposition of which it is true both that (a) there is some proposition *q* he believes such that *p* is at least as well confirmed, relative to *i*, as *q* is, and that (b) the set of the observations made is strictly serviceable, in the context of *i*, with regard to any problem characterized by $\{p, q, \ldots, z\}$, where $\{p, q, ..., z\}$ and i are as before. This raises fewer methodological difficulties than the original principle. Unfortunately, no comparative version of P is available, and so we cannot retire the quantitative concept of confirmation.

A more striking simplification is provided for. I have discussed the various sorts of issues separately, but it may have occurred to the reader that questions cannot properly be distinguished from problems and inquiries. A question may be said to focus on a one-membered set of basic alternatives. If the proposition in question is not a logical consequence of the information involved – and thus is not logically exhaustive, relative to the information – the set of alternatives characterizing that question is a problem-set, and the question itself is a problem. If the proposition in question does follow from the information, the set of alternatives is an inquiry-set, and the question is an inquiry. In the former case, E applies and requires simply that we believe p if and only if $c(p, i) \geqslant 1 - \varepsilon$, and if we agree to set $1 - \varepsilon = \omega$, this is exactly what S requires. (We can now see that if we did *not* set $1 - \varepsilon = \omega$, we would have a contradiction on our hands.) In the latter case, P applies and requires that we believe the proposition in question, which again is what S requires in this case. The functions of S are thus divided between E and P, and S becomes a subsidiary principle. Subjectivism is absorbed by empiricism and pragmatism.

The two principles that remain keep out of each other's way. E applies only where, relative to the information, the basic alternatives are not exhaustive. P applies only where they are exhaustive. Each principle is relevant to a situation to which the other is not. It follows that we cannot easily dispense with either of them. A comprehensive theory of rational belief will have to endorse them both, or endorse principles serving the same purposes.

To say that only one of our two principles can apply at any occasion is not to suggest that the other cannot come to apply in the sequel. If today we are committed to believing a proposition by E, we can come tomorrow to be committed to believing it by P, but only as a result of a change of issue. An issue is an organization of a quandary, and we are always free to restructure the quandaries in which we find ourselves. A problem can, with a change of perspective, give way to an inquiry, and vice versa. But insofar as we see things in one way, we do not see them in another. Insofar as one principle applies, the other consequently does not. In this sense, the principles are distinct, and we need them both.

Do we need any additional principles on this level of abstraction? This depends on whether there are epistemic issues of other sorts than those we have discussed. We have considered issues in which the basic alternatives are collectively exhaustive and issues in which they are not. But the only cases of these sorts we have considered are cases in which the basic alternatives are pairwise exclusive. This suggests that we look at situations in which the basic alternatives are *not* pairwise exclusive. This would be an odd sort of case. Since several of the basic alternatives might be true, we ought to be entitled to believe more than one. How then could this make for an issue? What would the quandary be? Unless I am missing some subtlety here, the principles we have considered cover the field.

V. SOME OVERARCHING PRINCIPLES

Each of the major principles developed above sets forth only what is distinctive of the position it represents. There are also principles that apply alike in each of the sorts of cases to which the special principles apply. The *Principle of Coherence* has already been cited. This is:

(*A1*) A person believes p only if $c(p, i) > c(\text{not-}p, i)$.

This principle is generally accepted, but not universally. (Levi, for one, rejects it.) It strikes me as sound. It is, at any rate, easy enough to accommodate. Its impact is simply to restrict us to values of ω greater than 0.5, and thus to values of ε and of η *less* than 0.5.

To say that *A1* seems sound is not to suggest that it must be adopted. We could make do without it. There would then be no restrictions on ω, and thus none on ε and on η, and this would allow a person to believe incompatible propositions, and to believe propositions incompatible with his information. A person might even find himself committed to some beliefs of this sort. We are likely to be predisposed against all this, but within our framework here we may be at ease on the subject. If any of the special principles we have endorsed did commit us to believing incompatible propositions, each of these commitments would hold in the context of a different issue. They would thus be insulated from one another, and could not conflict. As for believing propositions incompatible with our information, we could discredit this without adopting *A1*. We need only require that a person not believe every proposition that comes into question for him, or more directly, that he not believe any propositions incompatible with his information. The most that can be said for *A1* is that it rings true, and that no harm comes of adopting it. There are no compelling reasons for doing so.

A second overarching principle that has been proposed is the *Principle of Conjunctivity:*

(*A2*) If a person believes both p and q, he also believes p-and-q.

I have argued that *A2*, as it stands, is unacceptable, that counter-instances are available.[7] But the objection to *A2* goes deeper. We consider a person's commitments to be relative to the information he has. We ought, by the same token, to consider them relative to the issue he is raising. If a person believes p on information i, and believes q on i', we do not require that he believe p-and-q in either of these cases. If he believes p in the context of issue u, and believes q in the context of u', we ought likewise not to require that he believe p-and-q in either context.[8] Rationality demands at most that a person believe the conjunction of those propositions that he believes in the same context. We must therefore understand p and q in *A2* to be basic alternatives or alternations of basic alternatives in one issue. But this interpretation deprives *A2* of all force. Since neither of our two fundamental special principles entitles us to believe more than a single proposition in the context of any issue, the opportunity for violating *A2* no longer arises. A person who adopts *E* and *P* can adopt *A2* as a vacuous corollary.

A third proposed overarching principle is the *Principle of Deducibility:*

(*A3*) If a person believes p, and q is a deductive consequence of p, he also believes q,

where p is a basic alternative or an alternation of basic alternatives in the issue the person is facing. (If this qualification were not made, we would never be entitled *not* to believe a proposition in the context of one issue while believing it in the context of another.)

A3 is far too demanding. It requires that we believe an infinite number of propositions, many of them partly compounded of propositions not among our basic alternatives. (If p, q and r are our basic alternatives, and p is the proposition we believe, *A3* would commit us to believing not only *p*-or-*q* and *p*-or-*r*, but also *p*-or-*v*, *p*-or-*w*, *p*-or-*x*-or-*y*, etc.) This is excessive on several counts. Most obviously, an infinite number of beliefs may be more than any person can manage, in which case the dictum *ought implies can* precludes demanding this much. But the obligation to believe propositions not compounded of our basic alternatives is equally unacceptable. All such beliefs would introduce irrelevancies. Their adoption could in no way contribute to resolving the issues at hand. There is therefore no reason why we should acknowledge any commitments to them.

We might revise *A3*, requiring only that a person believe those deductive consequences of the propositions he believes that are compounded of his basic alternatives, or rather (to avoid the introduction of excluded options) only those that are alternations of his basic alternatives. But the revised *A3* would still be unacceptable. Our special principles commit us to believing the propositions that best resolve the issues we face. Once an issue has been settled, no purpose is served by believing any further propositions in the same context, and this no matter how these additional propositions are compounded. The best is as much as we need. Why then should we be committed to believing more?

But a more troublesome question comes to mind: what would such a commitment commit us to? What would a person be doing if, believing one of the alternatives of an issue, he believed also those alternations of the basic alternatives that follow deductively from it – if, say, believing *p*, he also believed *p*-or-*q*? If he neither believed *p* nor believed *q*, believing *p*-or-*q* would be a case of hedging. But since he does believe *p*, believing *p*-or-*q* must receive some other interpretation, and how it is to be interpreted – what psychological state it might be said to involve – is hard to say. It is in fact an open question whether such a belief could be maintained at all. (How would I go about believing that I was either Schick *or* Levi?) I do not know what to make of it.[9] Since the adoption of *A3* would commit us to frames of mind there seems to be no way of identifying, since it might indeed commit us to spurious postures, I think it best to do without it.

The deducibility requirement is none the more acceptable if it is built directly into some special principle of belief. *P′* incorporates a truncation of the revised version of *A3:* it requires us to believe all those alternations of our basic alternatives that are deductive consequences of the propositions we are committed to believing and have an expected epistemic utility equal to or greater than $1-\eta$. P_L picks up *A3* as it stands. As I see it this does not do anything for deducibility. It serves only to discredit these principles. (Hence my preference for *P* over both of them.)

Let me remark that the rejection of *A3* does not make for libertarianism. It does not allow anyone to refuse to believe the deductive consequences of his beliefs. For if a person is to reject a deducible proposition, it must first come into question for him. (In which case, I take it, *he* understands the option.) And once the proposition is in question, he is committed to believing it by *S*, for the degree of confirmation of a proposition is at least as great as that of any from which it follows. He may not be committed to believing a deducible proposition on his own initiative, but he cannot properly refuse to believe it after it has come to his attention. (It may be that this is all that is intended by those who would endorse *A3*.)

In sum, of the three most familiar overarching principles, only one, the Principle of Coherence, merits consideration. In the context of our analysis here, the Principle of Conjunctivity, when plausibly interpreted, is redundant, and the Principle of Deducibility is best dropped altogether.

Rutgers University, New Brunswick, N.J.

BIBLIOGRAPHY

[1] Hempel, Carl G., 'Inductive Inconsistencies', *Synthese* **12** (1960) 439–69.

[2] Hempel, Carl G., 'Deductive-Nomological vs. Statistical Explanation', in *Minnesota Studies in the Philosophy of Science*, vol. 3 (ed. by H. Feigl and G. Maxwell), Minneapolis 1962.

[3] Hilpinen, Risto, 'Rules of Acceptance and Inductive Logic', *Acta Philosophica Fennica* **22** (1968).

[4] Hintikka, Jaakko, 'Towards a Theory of Inductive Generalization', in *Logic, Methodology and Philosophy of Science* (ed. by Y. Bar-Hillel), Amsterdam 1966.

[5] Hintikka, Jaakko, 'On a Combined System of Inductive Logic', *Acta Philosophica Fennica* **18** (1965) 21–30.

[6] Hintikka, Jaakko and Hilpinen, Risto, 'Knowledge, Acceptance and Inductive Logic', in *Aspects of Inductive Logic* (ed. by J. Hintikka and P. Suppes), Amsterdam 1966.

[7] Hintikka, Jaakko and Pietarinen, Juhani, 'Semantic Information and Inductive Logic', in *Aspects of Inductive Logic* (ed. by J. Hintikka and P. Suppes), Amsterdam 1966.

[8] Humphrey, George, *Thinking*, New York 1963.

[9] Köhler, Wolfgang, *Gestalt Psychology*, New York 1959.

[10] Kyburg, Henry E., Jr., *Probability and the Logic of Rational Belief*, Middletown, Conn., 1961.

[11] Levi, Isaac, *Gambling With Truth*, New York 1967.
[12] Levi, Isaac, 'On the Seriousness of Mistakes', *Philosophy of Science* **29** (1962) 47–65.
[13] Miller, George A., 'The Magical Number Seven, Plus or Minus Two: Some Limits on our Capacity for Processing Information', *Psychological Review* **63** (1956) 81–97.
[14] Morgenbesser, Sidney, 'Goodman on the Ravens', *Journal of Philosophy* **59** (1962) 493–5.
[15] Scheffler, Israel, *The Anatomy of Inquiry*, New York 1963.
[16] Schick, Frederic, 'Consistency', *Philosophical Review* **75** (1966) 467–95.
[17] Schick, Frederic, 'Consistency and Rationality', *Journal of Philosophy* **60** (1963) 5–19.

REFERENCES

1 The restriction of its scope to the field of questions is the respect in which *S* moves furthest from Kyburg's position in [10], from which it is descended. (I comment on Kyburg's analysis in [17].) A closer ancestor is what I call the Principle of Order in my [16], but this too is unrestricted in scope.

2 The only-if clause of *S* is vacuous. Its sole function is to facilitate comparisons with the principles to be introduced in subsequent sections.

3 See [6]. Unlike E_H, Hintikka and Hilpinen's principle is intended to apply only where generalizations are being considered for belief. The extension of its scope to include singular propositions is thought to encounter difficulties with conjunctivity. (I discuss conjunctivity in Section V.) I have also made some changes to facilitate the comparison of empiricism with subjectivism and with pragmatism, but these changes are trifling.

4 Hintikka and Hilpinen go one step further so as to make n_0 independent of the number of Q-predicates exemplified in the sample ([6], pp. 10–1), but this independence strikes me as implausible. (Hilpinen himself has had some second thoughts on this matter; see [3], p. 67.) I should add that my formulation of Hintikka and Hilpinen's position reflects my replacement of their '$>$' by '$\geqslant$' in E_H and my deletion of their requirement that ε be less than 0.5.

5 This concept of serviceability is suggested by Morgenbesser [14] and by Scheffler [15], pp. 286–91. Morgenbesser and Scheffler discuss the serviceability of observation-*reports*, which is indeed syntactically more appropriate than my terminology above.

6 Levi himself goes further afield in [12]. For other maximizing analyses, see Hempel [1] and [2], Hintikka and Pietarinen [7] and Hilpinen [3].

7 See [16], pp. 479–80. These counter-instances derive from Kyburg [10].

8 This argument is taken from Levi [11].

9 I may, in preceeding sections, have given the impression that I did – e.g. "believing this alternation and no sub-alternation maintains the *status quo*". But this served only to anticipate misguided objections.

MARSHALL SWAIN

THE CONSISTENCY OF RATIONAL BELIEF*

When are a man's beliefs rational, and when are they irrational? A systematic answer to these questions constitutes a theory of the ethics of belief; that is, a theory which tells us what a man is obligated, permitted, and forbidden to believe, from the point of view of rationality. A theory of the rationality of belief will consist in large part of a system of rules which determine, in a given cognitive situation, what a man's doxastic commitments ought to be. The most general form of a theory of rational belief will incorporate a set of rules that tell us what our doxastic commitments are once it has been determined which propositions, in a given cognitive situation, are justified, evident, reasonable, unreasonable, and so forth. A general theory of this sort becomes more exact when it is supplemented with a set of rules that also determine, in a given situation, which propositions *are* justified, evident, etc. This latter body of rules will generally incorporate, but not wholly consist of, criteria of inductive support. In addition to criteria of inductive support, such rules will generally incorporate criteria for determining which nonexperiential propositions, if any, are justified independently of the need for inductive support, and which experiential propositions are justified but not justified by any other propositions. These criteria might serve as explications of the general concepts Justified, Evident, Reasonable, and so forth. Thus, the rules incorporated in a general theory of rational belief can be used to formulate a partial criterion for adequate explications of these concepts; if a proposed explication violates the rules of a general theory, then it is inadequate. In this paper, I shall construct a general theory of rational belief which hopefully will serve at least this purpose.

I. SOME GROUNDWORK

A. *The Concept of Epistemic Consistency*

I propose that a general theory of doxastic commitment can profitably be construed as a theory about the consistency of sets of expressions which contain the following epistemic locutions (among others)[1]:

"*a* believes that *p*"
"*p* is reasonable for *a*"
"*p* is unreasonable for *a*"
"*p* is evident for *a*".

These locutions are representative of two basic categories of epistemic locutions. On the one hand, there are expressions such as "*a* believes that *p*", "*a* knows that *p*", "*a* remembers that *p*", which ascribe epistemic *states* to the person referred to by '*a*'. On the other hand, there are expressions such as "*p* is evident for *a*", "*p* is reasonable for *a*", and "*p* is unreasonable for *a*", which ascribe an epistemic value to the proposition *p* for the person *a*. The former type of expression will be called an *epistemic state ascription*, and the latter an *epistemic value ascription*.

I shall take the relationship of *epistemic consistency* as the primary relationship to be established among these locutions. Perhaps the easiest way to understand this relationship is to consider various cases of epistemic inconsistency. Thus, suppose "*a* believes that *p*" and "*a* believes that not-*p*" are both true of the person referred to by '*a*'. Or, suppose *a* believes that *p* and believes that not-*q*, where *p* entails *q*. In both of these cases, we would say that there is something *wrong* with the epistemic state of affairs described by these locutions. If these expressions are true of the person referred to by '*a*', then that person falls short of 'epistemic perfection'; there is, in these cases, a certain line of criticism that we could bring to bear on *a* such that if *a* persisted in believing the things that he does we would dismiss him as irrational. With respect to the examples given, the criticism would be logical in nature; we could show *a* that the propositions he believes are logically inconsistent.

Criticisms based on considerations of logical consistency are not the only kind relevant to epistemic consistency. For example, suppose *a* believes that *p* but fails to believe *q* (where *p* entails *q*) even though he does not believe "not-*q*". If so, then *a*'s beliefs may very well be logically consistent; however, we could criticize *a* on the grounds that the set of propositions he believes is deductively incomplete. There is something implicitly irrational about the structure of *a*'s beliefs; accordingly, the set of sentences describing that structure is epistemically inconsistent.

The examples of epistemic inconsistency just described involve only epistemic state ascriptions. There are also instances of epistemic inconsistency among state and value ascriptions. For example, the locutions "*p* is evident for *a*" and "*a* believes that not-*p*" are epistemically inconsistent; the person referred to by '*a*' would fall short of perfect rationality if these were both true of him.

On the basis of these examples, we may informally characterize the concept of epistemic consistency as follows: A set of sentences is epistemically inconsistent if and only if not all of its members could simultaneously be true of an epistemically perfect entity; a set of sentences is consistent just in case it is not inconsistent.

In addition to the above, there are examples of inconsistency among value ascriptions. Thus, "*p* is evident for *a*" and "not-*p* is evident for *a*" describe a situation which we would want to call inconsistent. But there is an important difference between this case of inconsistency and the previous cases; namely, in the case just described we would hesitate to say that the inconsistent locutions *could* both be true of the person referred to by '*a*', while in the previous cases there is no serious problem about that assumption. Any adequate explication of the concepts Evident and Believes should, for example, leave open the possibility that *a* fails to believe *p* when *p* is evident for him, but should not allow that a proposition and its denial are both evident for a person (at one and the same time). Thus, we might want to say that "*p* is evident for *a*" and "not-*p* is evident for *a*" are implicitly *logically inconsistent*, and that this logical inconsistency can be brought out in the open by considering the explication of the concept Evident. This is no problem for the concept of epistemic consistency, however, for two expressions are epistemically consistent only if they are logically consistent. The informal characterization of epistemic inconsistency given above is rather trivially satisfied in this example, for if two expressions are logically inconsistent then they cannot simultaneously be true of *any* entity, including an epistemically perfect one.

These considerations show that epistemic consistency is a much stronger concept than logical consistency; logical consistency is a necessary but not a sufficient condition for epistemic consistency.

Moreover, the concept of epistemic consistency is obviously highly idealized, in the sense that the set of sentences describing the epistemic state of affairs of any ordinary mortal will undoubtedly turn out to be inconsistent. This should not be considered a defect in the theory, however. To say that a man's beliefs are epistemically inconsistent is not necessarily to say that he is blameworthy; to say that a man falls short of epistemic perfection does not imply that he is irrational in any pragmatically meaningful sense of that term. To echo a remark of Frederic Schick's, it is no more essential to the adequacy of our theory that anyone be consistent than it is for ethics that anyone be a saint.[2]

B. *Clarification of the Epistemic Concepts*

This comparison with ethical theory is an instructive one, for there is a considerable analogy between our epistemic evaluation terms and moral terms. Thus, as we shall use the term here, 'evident' is analogous to the moral term 'obligatory'; to say that a proposition is evident for a man is analogous to saying that an action is obligatory for him. Similarly, 'reasonable' is analogous to 'permitted', and 'unreasonable' to 'forbidden'. Finally, believing that a proposition is true is analogous to performing an action.[3]

Just as in moral contexts we speak of certain right-making characteristics in terms of which an action can be classified as obligatory, permitted, or forbidden, we can, in cognitive contexts, point to certain characteristics in terms of which a proposition can be classified as evident, reasonable, or unreasonable. To say that a proposition is evident for a man is to say that it has a certain property, and it can come to have this property in any of a number of ways. Some philosophers have maintained that there is, for example, a set of propositions each member of which is *directly* evident for a given person; this set might include basic propositions, phenomenal reports, first-person introspective reports of sensations, pains, and so forth. There is very little agreement (to put it mildly) about what the members of this set are, and indeed very little agreement about whether there is such a set of propositions.

Sometimes it is suggested that there is also a set of propositions each member of which is *self-evident* for a given person; this set might include a priori truths, necessary truths, logical truths, and so forth. Again, there is disagreement about just what members this set has, and whether there is such a set.

In addition, there is a set of propositions that are neither directly evident nor self-evident for a man, but are evident in virtue of their relations to other propositions that are evident for him. In a given typical cognitive situation, a man will have a body of evidence, and there will be a set of propositions such that relative to that body of evidence each member of this set has the property of being evident. The relationship that holds between the evidence and given members of this set might be deductive or inductive. Whenever a relationship holds between a man's body of evidence *e* and some proposition *p* such that *p* is evident relative to *e*, then I shall say that *e indicates p*. The problem of defining this relationship is surely one of the central tasks of inductive logic.

In terms of these reflections, the following is suggested as a preliminary schema for an explication of the concept Evident:

(I) *p* is evident for *a* if and only if
either (1) *p* is directly evident for *a*;
or (2) *p* is self-evident for *a*;
or (3) *p* is indicated by the set of propositions which satisfies (1) and (2);
or (4) *p* is indicated by the set of propositions which satisfies (1) or (2) or (3);
or etc...

The concepts Reasonable and Unreasonable are, as has been pointed out, analogous respectively to the moral concepts Permitted and Forbidden. Clearly, if a proposition is evident for a man, then it is reasonable for him and its denial is unreasonable. However, there will be cases in which a proposition is reasonable but not evident. These will be situations in which the proposition *p* has a relationship to *a*'s evidence which is weaker than the relationship that I have called 'indication'. Whenever a proposition *p* is related to *a*'s evidence *e* in such a way that *p* is reasonable for *a*, I shall say that *e sustains p*. One straightforward way of explicating the relation of sustenance is: *e* sustains *p* if and only if the probability of *p*, given *e*, is equal to or greater than *k* (where *k* is a value greater than .5 and less than or equal to 1). I have left the value *k* open because I cannot think of any intelligent way of picking one. Whether this 'simple' interpretation of the sustenance relation will suffice must remain temporarily problematic; it will, in any event, suffice for my discussion. The following preliminary schema for the explication of the concept Reasonable is thus suggested:

(II) *p* is reasonable for *a* if and only if
either (1) *p* is evident for *a*
or (2) The set of propositions *e* such that each member of *e* satisfies (1) is such that *e* sustains *p*.

Then, given this schema, we can easily characterize the concept Unreasonable:

(III) *p* is unreasonable for *a* if and only if it is not the case that *p* is reasonable for *a*.

The schematic explications of our epistemic concepts suggested above allow considerable latitude of interpretation. They can, hopefully, be satisfied by any of a number of explicata of the concepts Directly Evident, Self-evident, Indicated, Sustained, and related concepts. It must also be recognized that these schematic explications might themselves stand in need of revision; their presentation here is only intended to help clarify the meanings of our epistemic locutions.

II. THE THEORY

The approach to epistemic consistency to be developed in this paper is intimately based upon the work done in epistemic logic by Hintikka.[4] Indeed, the reader who is familiar with Hintikka's work will find many similarities of method and terminology between Hintikka's theory and the one presented here; this indebtedness will not always be signaled. Despite the similarities, however, there are also important dissimilarities. For example, while the concept of epistemic consistency is intended to be analogous to Hintikka's concept of defensibility, the informal as well as the formal accounts of the former concept given here are significantly different from Hintikka's account of the latter concept.[5] But even so, my theory can be considered an attempt to build upon Hintikka's theory; his theory deals primarily with the consistency (defensibility) of the epistemic *state* ascriptions "*a* knows that *p*" and "*a* believes that *p*", while my theory attempts to include an account of epistemic value ascriptions.

A. *Consistency and Epistemic Universes*

If, for a given person *a*, we were to list all of his epistemic state and value ascriptions at a given time, we would have a complete description of his epistemic state of affairs at that time; this set of sentences shall be called the *epistemic framework* of the person *a*. Thus, if *a* believes that *p*, then "*a* believes that *p*" will be a member of his epistemic framework, and if *p* is evident for *a*, then "*p* is evident for *a*" will be a member of his framework. In addition to epistemic frameworks, we can single out the set of propositions such that each member of the set is the object of one of *a*'s epistemic states. We select from this set all the members that are *true*; for example, if *a* believes that *p* and *p* is in fact true, then *p* is a member of this set. We then take the set of sentences such that for every member ϕ either (1) ϕ is an epistemic state ascription for *a*, or (2) ϕ is an epistemic value ascription for *a*, or (3) ϕ is the object of one of *a*'s epistemic states and is true, or (4) ϕ is logically deducible from one of the sentences satisfying (1), (2), or (3). This set of sentences will be called the *epistemic universe* of the person *a*, and will be represented formally by the symbol 'U*a*'.

An epistemic universe U*a* will be assumed to have a time factor built into it; thus, 'U*a*' will represent the epistemic state of affairs of *a* at a given time, or over an interval of time during which *a*'s beliefs, etc., do not change. Any given U*a* will thus represent a complete cross-section of what we might call *a*'s 'epistemic biography'.

Epistemic consistency and inconsistency will be defined in the primary sense as properties of epistemic universes. A set of principles, in the form of rules of admissibility, will be stipulated such that an epistemic universe is consistent if and only if it satisfies these rules. It will be epistemically inconsistent, or imperfect, otherwise. To say that these principles can also serve to formulate a criterion of adequacy for any proposed explication of the evaluative concepts involved is to say that any such explication must not yield, in a given cognitive situation, results that are epistemically inconsistent.

B. *Logical Consistency and Deductive Closure*

I shall now introduce the rules referred to above which govern the admissibility of sentences into consistent epistemic universes. The first of these rules are fairly uncontroversial; they are designed to guarantee that epistemic universes are logically consistent and deductively closed.

(C.cons) If $p1 \in Ua$ and $p2 \in Ua$ and...and $pn \in Ua$, then if the conjunction of q with $p1$ and $p2$ and ... and pn entails a contradiction, it is not the case that $q \in Ua$.

(C.ded) If $p1 \in Ua$ and $p2 \in Ua$ and...and $pn \in Ua$, then if the conjunction of $p1$ and $p2$ and...and pn entails q then $q \in Ua$.

(C.cons) entails the following rule, which will be of some use later:

(C.$\sim$) If $p \in Ua$, then it is not the case that "$\sim p$" $\in Ua$.

The rule (C.cons) formulates what Kyburg has called a 'strong' principle of consistency[6], which is to be contrasted with the 'weak' consistency principle (C.$\sim$). A set of sentences might satisfy the rule (C.$\sim$) without satisfying (C.cons). In some contexts (see, for example, Section III of this paper) the weak principle is clearly the desirable one; however, in characterizing the general contents of a consistent epistemic universe, I believe that the stronger principle is required.

(C.cons) is, in any event, implied by (C.ded) in conjunction with (C.$\sim$). The rule (C.ded) formulates a 'strong' deductive closure requirement, to be contrasted with a 'weak' closure principle such as "If $p \in Ua$, and if p entails q, then $q \in Ua$". A set of sentences might satisfy the weak principle without satisfying the strong one. Again, in characterizing the general contents of a consistent epistemic universe, the strong principle is the desirable one.

It is important to realize that the rules (C.cons) and (C.ded) apply to epistemic universes as characterized in Section II.A above. We will ultimately be concerned with certain sets that can be generated from a given epistemic universe; for example, the set of sentences R such that all and only the members of R are reasonable for a, relative to the Ua in question. Similarly, we will be concerned with the sets of sentences that are believed, evident, unreasonable, and so forth. At this stage, the rules we have adopted tell us very little about the restrictions to be placed on these sets of sentences. It would be a mistake to assume that these sets will satisfy the general restrictions on the contents of a perfect epistemic universe; indeed, some of them will not.

Our problem now is to determine what additional rules we will need over and above these rules; what rules are required for the admissibility of epistemic state and value ascriptions into a perfect epistemic universe?

C. *Consistency of Belief*

We shall use 'B*ap*' as the formal counterpart for '*a* believes that *p*'. In addition, we shall use another epistemic value concept; namely, 'It is compatible with everything *a* believes that *p*', represented formally by 'C*ap*'. This expression is to be interpreted in a perfectly straightforward way; 'compatible' here means logically compatible. Thus, to say that *p* is compatible with everything you believe is to say that the conjunction of *p* with all of the propositions that you believe is logically consistent. This expression will be of great use in formulating our conditions.

In order to capture the requirements that must be placed on epistemic universes if they are to be consistent, we shall have to consider descriptions of possible states of affairs that might have obtained instead of the state of affairs described by a given epistemic universe. Such descriptions will be called "alternatives" to the epistemic universe in question. These will be represented formally by 'U*a**', 'U*a***', etc. In particular, we shall be concerned with a special sort of alternative, called a "doxastic alternative". A doxastic alternative is defined as follows[7]:

> The epistemic universe U*a** is a doxastic alternative to the epistemic universe U*a* if and only if (i) U*a** describes a situation in which *a* believes everything that he believes in the situation described by U*a* and (ii) everything that *a* believes in the situation described by U*a* is true in the situation described by U*a** and (iii) U*a** satisfies all of the C-rules satisfied by U*a*.

Intuitively, a doxastic alternative to an epistemic universe is a situation in which everything you believe comes to be true. Notice that, as defined, a doxastic alternative does not limit *a* to believing only what he believes in the situation described by U*a*; notice, also, that U*a* might be a doxastic alternative to itself.

These concepts provide us with a way of partially characterizing the consistency of epistemic universes. For example, in order for *p* to be compatible with everything that you believe, there must be at least one doxastic alternative to the universe describing your beliefs such that *p* is true in that alternative. This is captured in the following rule:

(C.C*) If "Cap" $\in$ Ua, then there is at least one doxastic alternative Ua* to Ua such that $p \in$ Ua*.

(C.C*) tells us, again, that if p is to be compatible with all that you believe, then there must be at least one possible world (perhaps even this actual world) such that in that world all you believe in this actual world is in fact true and p is also true. The next two rules are easily justified by reference to the definition of a doxastic alternative.

(C.BB*) If "Bap" $\in$ Ua, then for every doxastic alternative Ua* to Ua, "Bap" $\in$ Ua*.

(C.B*) If "Bap" $\in$ Ua, then for every doxastic alternative Ua* to Ua, $p \in$ Ua*.

In addition, the following rule seems acceptable.

(C.$\sim$C) If "$\sim$ Cap" $\in$ Ua, then "B$a \sim p$" $\in$ Ua.

This rule says that if p is incompatible with what you believe, then you ought to believe "$\sim p$". In effect, this rule commits one to believing all propositions entailed by any proposition that he believes. From (C.$\sim$C) we can derive

(C.$\sim$B) If "$\sim$ Bap" $\in$ Ua, then "C$a \sim p$" $\in$ Ua.

The rules presented thus far suffice to guarantee the logical consistency of belief; that is, they guarantee that the set $B = \{p1, p2, \ldots, pn\}$ of sentences believed by a (at a given time) is logically consistent. In addition, the rules guarantee that B is deductively closed in the strong sense referred to in Section II.B above. Since some of the members of B might be false, there is no guarantee that B is a subset of Ua; however, the restrictions on Ua guarantee that B is consistent and closed. These results, and others as well, are demonstrated in the Appendix to this paper.

If a set of beliefs is inconsistent according to the rules developed thus far, then it cannot be rational. However, given that a set of beliefs satisfies the above rules, it remains open whether that set of beliefs is rational, or whether the epistemic universe in question is consistent. I might, for example, believe that the earth is flat and that the sun revolves around it, and provided that my other beliefs are stipulated accordingly, my beliefs might very well satisfy the rules presented thus far. But of course these beliefs are irrational, given the evidence that I have.

Whether a man's beliefs are rational depends in large part on the epistemic status of those beliefs; and this is, in turn, a question of epistemic value. We must, therefore, expand our theory to account for epistemic value.

D. *Epistemic Value Ascriptions*

We shall need, in addition to the rules governing belief ascriptions, rules that govern the admissibility of epistemic value ascriptions into consistent epistemic universes. The expression '*p* is evident for *a*' will be represented formally by 'E*ap*'; '*p* is reasonable for *a*' will be represented by 'R*ap*'; and '*p* is unreasonable for *a*' by 'U*ap*'. In addition, we shall rely on the expression 'it is compatible with everything that is evident for *a* that *p*' which is represented formally by 'L*ap*'. Just as we developed the notion of a doxastic alternative to an epistemic universe, we shall also develop the notion of an *evidential alternative* to an epistemic universe. Intuitively, an evidential alternative U*a** to an epistemic universe U*a* will be a state of affairs in which everything that is evident for *a* in U*a* is evident for *a* and true. Such alternatives may be characterized as follows:

> U*a** is an evidential alternative to the epistemic universe U*a* if and only if (i) U*a** describes a situation in which everything is evident for *a* that is evident for *a* in the situation described by U*a* and (ii) everything that is evident for *a* in U*a* is true in the situation described by U*a** and (iii) U*a** satisfies all the C-rules satisfied by U*a*.

The rules governing doxastic alternatives that were introduced earlier have analogues for evidential alternatives. Thus, the following are adopted:

(C.L*) If "L*ap*" $\in$ U*a*, then there is at least one evidential alternative U*a** to U*a* such that *p* $\in$ U*a**.

(C.EE*) If "E*ap*" $\in$ U*a*, then for every evidential alternative U*a** to U*a*, "E*ap*" $\in$ U*a**.

(C.E*) If "E*ap*" $\in$ U*a*, then for every evidential alternative U*a** to U*a*, *p* $\in$ U*a**.

In addition, we adopt the following rules governing the admissibility of epistemic value ascriptions into perfect universes:

(C.∼L) If "∼ L*ap*" ∈ U*a*, then "E*a* ∼ *p*" ∈ U*a*.
(C.E) If "E*ap*" ∈ U*a*, then "∼ E*a* ∼ *p*" ∈ U*a*.
(C.Rded) If "R*ap*" ∈ U*a* and *p* entails *q*, then "R*aq*" ∈ U*a*.
(C.E∼R) If "E*ap*" ∈ U*a*, then "∼ R*a* ∼ *p*" ∈ U*a*.
(C.U∼E) If "U*ap*" ∈ U*a*, then "∼ E*ap*" ∈ U*a*.
(C.R∼U) "R*ap*" ∈ U*a* if and only if "∼ U*ap*" ∈ U*a*.

From (C.∼L) we can derive

(C.∼E) If "∼ E*ap*" ∈ U*a*, then "L*a* ∼ *p*" ∈ U*a*.

These rules suffice to guarantee the epistemic consistency of sets that contain our epistemic *value* ascriptions.

E. *The Consistency of Rational Belief*

In addition to these rules governing belief and value ascriptions, we need rules *linking* these expressions. Such rules will tell us whether it is rational to believe, disbelieve, or suspend belief about *p* given the epistemic value that *p* has. Only two additional rules are necessary:

(C.EB) If "E*ap*" ∈ U*a*, then "B*ap*" ∈ U*a*.
(C.BR) If "B*ap*" ∈ U*a*, then "R*ap*" ∈ U*a*.

The addition of (C.EB) and (C.BR) completes the list of rules that I shall adopt. These rules suffice to guarantee that *a*'s beliefs are consistent *and* rational.

There are rules other than these that are tempting. For example, there is intuitive appeal to the suggestion that "E*a*E*ap*" implies "E*ap*". However, this cannot be proven given the extant C-rules; that is, {"E*a*E*ap*", "∼ E*ap*"} is consistent according to these rules. We could establish this principle by adding a condition that guarantees the density of the relation of evidential alternativeness. This rule would be:

(C.dens) If U*a** is an evidential alternative to U*a*, then there is at least one evidential alternative U*a*** such that U*a*** is an evidential alternative to U*a* and U*a** is an evidential alternative to U*a***.

However, doubt is cast upon the validity of the principle that "E*a*E*ap*" implies "E*ap*" by the dubiousness of the similar principle that "E*ap*" implies *p*. Adopting (C.dens) will *not* enable us to prove this latter principle, but its dubiousness casts doubt on the former principle. Thus, I shall not at present adopt (C.dens).[8]

In the Appendix to this paper the reader will find proofs of a number of interesting theorems. That the rules presented above represent an intuitively plausible theory about the logical structure of rational belief will hopefully be demonstrated in what follows.

III. AN APPLICATION OF THE THEORY; THE LOTTERY PARADOX

If the principles of epistemic consistency that have been suggested are basically correct, then they ought to provide guidelines for the solution of problems concerning rational belief. In this section I shall discuss one such problem, and endeavor to show that this theory provides an intuitively plausible solution to it. In the course of this discussion a number of interesting theorems will be brought to light.

One of the more recalcitrant problems that has been under discussion lately is the paradox of the lottery, suggested by Kyburg.[9] This paradox arises because of a conflict among seemingly acceptable canons of rational belief. I shall develop the problem in a way that lends itself to discussion in terms of the theory that has been presented above; no doubt there are other ways of formulating it.

Imagine a fair lottery in which there are 1000 tickets, one of which will be drawn. We may suppose that each of the tickets has been marked uniquely with one of the numerals '1' through '1000'. Let '*p*1' stand for 'Ticket number one will be drawn', '*p*2' for 'Ticket number two will be drawn', and so forth. In addition, let '*q*' stand for the disjunction of the statements '*p*1' through '*p*1000'. Suppose *a* knows that the lottery is fair, and that exactly one ticket will be drawn. Given *a*'s evidence, each statement '$\sim p1$' through '$\sim p1000$' has a probability of .999; moreover, '*q*' is entailed by this evidence, and thus has a probability of 1.

Each of the following canons of rationality has found support in the literature[10]:

(1) If *a*'s evidence entails *p*, then *p* is justified for *a*.

(2) If *p* is very highly probable, relative to *a*'s evidence, then *p* is justified for *a*.

(3) If *p* is justified for *a*, then if *a* is fully rational then *a* believes that *p*.

(4) If *a* believes that *p* and *a* believes that *q*, then if *a* is fully rational then *a* believes that *p* and *q*.

Each of these principles does, indeed, have a certain intuitive plausibility. However, in the lottery case these principles yield unhappy results. Given (1) it follows that

(i) *q* is justified for *a*.

Given (2) it follows that

(ii) $\sim p1$ is justified for *a* and $\sim p2$ is justified for *a* and ... and $\sim p1000$ is justified for *a*.

Given (3), (i), and (ii), it follows that

(iii) If *a* is fully rational, then *a* believes that $\sim p1$ and *a* believes that $\sim p2$ and ... and *a* believes that $\sim p1000$ and *a* believes that *q*.

But given (4) and (iii), it follows that

(iv) If *a* is fully rational, then *a* believes $\sim p1$ and $\sim p2$ and ... and $\sim p1000$ and *q*.

However, the conjunction referred to in the consequent of (iv) is logically inconsistent. This violates the surely acceptable principle that a fully rational being would not believe a contradictory statement. That is the paradox.

How can we avoid this paradoxical result? One suggested solution is to deny the closure condition for rational belief formulated in (4).[11] This would block the move from (iii) to (iv), thus avoiding the paradoxical result. However, denying (4) runs into direct conflict with the principle

(C.$\sim$B) If "$\sim Bap$" $\in$ U*a*, then "$Ca \sim p$" $\in$ U*a*.

Given the informal account of the property of consistency that these rules preserve, (C.$\sim$B) yields the following:

(5) If *a* is fully rational, then if $\sim$(*h* and *i*) is incompatible with everything that *a* believes then *a* believes that *h* and *i*.

Now to deny (4) is to allow that the following is *possible*:

(6) *a* is fully rational and *a* believes that *h* and *a* believes that *i* and *a* does not believe that *h* and *i*.

But (5) and (6) entail

(7) $\sim$(*h* and *i*) is compatible with everything that *a* believes

However, (6) also entails

(8) *a* believes that *h* and *a* believes that *i*.

And, of course, *h*, *i*, and $\sim$(*h* and *i*) are incompatible. So, (7) is false and the situation described in (6) is not in fact possible within the framework of my theory of rational belief. This suggested solution to the paradox must then be rejected if that theory is adopted.

Another solution can be developed in accordance with my theory, however. This solution rests on an ambiguity in the term 'justified' which occurs in (1)–(3). This term can be taken to mean 'evident' or to mean 'reasonable'. If we take it to mean 'evident', then (1) is surely acceptable; and, since every evident proposition is also reasonable, (1) is acceptable if we replace 'justified' with 'reasonable'. So, (1) can be accepted in either of the following variations:

(1′) If *a*'s evidence entails *p*, then *p* is evident for *a*.
(1″) If *a*'s evidence entails *p*, then *p* is reasonable for *a*.

However, principle (2) is acceptable only if we replace 'justified' with 'reasonable'. This is easily proven. If we replace 'justified' with 'evident' in (2), then we will encounter difficulties with the following principle:

(9) If *p* is evident for *a* and *q* is evident for *a*, then "*p* & *q*" is evident for *a*.

The principle (9) formulates a conjunctive closure condition for the set *E* of propositions that are evident for a man. This closure condition for *E* can be formally demonstrated within my theory. To prove that there is a closure condition for the set *E*, relative to a given U*a*, it suffices to prove that the set {"E*ap* & E*aq*", "$\sim$E*a*(*p* & *q*)"} cannot be imbedded in U*a* without violating the C-rules. To prove this we must first assume that "E*ap* &

E*aq*" as well as "$\sim$E*a*(*p* & *q*)" are members of U*a*:

(a) "E*ap* & E*aq*" $\in$ U*a* (assumption)
(b) "$\sim$ E*a*(*p* & *q*)" $\in$ U*a* (counterassumption).

We then show that the assumptions (a) and (b) lead, via a reductio ad absurdum, to a violation of (C.cons). From (b) and (C.$\sim$E) we derive

(c) "L*a* $\sim$ (*p* & *q*)" $\in$ U*a*.

Then, from (c) and (C.L*) we derive

(d) "$\sim$ (*p* & *q*)" $\in$ U*a**, where U*a** is an (i.e., some) evidential alternative to U*a*.

But from (a), (C.ded), and (C.E*) we can derive

(e) *p* $\in$ U*a**;
(f) *q* $\in$ U*a**.

But (d), (e), and (f) violate (C.cons), since *p*, *q*, and "$\sim$ (*p* & *q*)" conjoined entail a contradiction. Thus, our rules will not allow that "E*ap* & E*aq*" and "$\sim$E*a*(*p* & *q*)" be simultaneously true of *a*, and this establishes the closure condition (9) for the set *E*.

Now if we replace 'justified' with 'evident' in (2), then the resulting

(2′) If *p* is highly probable, relative to *a*'s evidence, then *p* is evident for *a*,

in conjunction with (9) leads to unhappy results. A particularly appropriate example is the lottery situation; for, in this case (9) and (2′) would imply that the conjunction of '$\sim$*p*1' through '$\sim$*p*1000' is evident for *a*. But this conjunction has zero probability relative to *a*'s evidence, and thus is surely not evident for him. Thus we must reject (2′) within the framework of my theory. On the other hand, if we replace 'justified' with 'reasonable', we get

(2″) If *p* is very highly probable, relative to *a*'s evidence, then *p* is reasonable for *a*.

(2″) will not lead to difficulties similar to those for (2′) because there *is no conjunctive closure condition* on the set *R* of propositions that are reasonable for a man. That there is no analogue to (9) for *R* is proven by showing that the set {"Rap & Raq", "~Ra(p & q)"} is consistently imbeddable in some Ua. This is the same as showing that there is at least one Ua such that the set consisting of Ua + the alternatives accessible from Ua (with respect to a) satisfies the rules.[12] The following will do:

Ua = {"Rap & Raq", "~Ra(p & q)", "Rap", "Raq", "Ua(p & q)", "~Ba(p & q)", "~Ea(p & q)", "Ca~(p & q)", "La~(p & q)", + whatever is required by (C.ded), (C.Rded)}
Ua* = {"~(p & q)", + whatever is required by (C.ded)}.

If we let Ua* be both a doxastic and an evidential alternative to Ua, and assume that there are no other alternatives, and that Ua and Ua* have only the members listed, then the set {Ua, Ua*} suffices to prove that {"Rap & Raq", "~Ra(p & q)"} is consistent, and thus to prove that closure fails for the set R. Thus, (2″) is acceptable, but (2′) is not.

Principle (3), on the other hand, is acceptable if we replace 'justified' with 'evident', but not if we replace it with 'reasonable'. This is easily verified by reference to (C.EB) and (C.BR). Thus we can accept

(3′) If p is evident for a, then if a is fully rational then a believes that p.

but should reject the analogue for 'reasonable'.

The acceptable versions of (1)–(4) are, then: (1′), (1″), (2″), (3′), and (4). Given these principles, the paradoxical results arrived at previously will no longer follow. Instead, we get the following results in the lottery case:

(v) q is evident for a
(vi) a believes that q
(vii) ~p1 is reasonable for a and ~p2 is reasonable for a and ... and ~p1000 is reasonable for a.

Lacking a closure condition for the set R we cannot conclude from (vii) that the conjunction of the propositions '~p1' through '~p1000' is reasonable for a, and lacking an analogue to (3′) we cannot conclude that a believes any of these propositions.

If we adopt something like the preliminary explication of 'reasonable' in Section II.B, however, then it is acceptable within the boundaries of my theory that a does in fact believe any one of the negative propositions;

and, indeed, *a* can consistently believe any of a number of conjunctions of these propositions. The range of propositions other than *q* which *a* can consistently believe is determined by the value assigned to *k* in our schematic explication of 'reasonable'. In general, *a* can consistently believe any proposition provided that it has a probability greater than or equal to *k* (relative to his evidence); and *a* can consistently believe any set of propositions provided that the conjunction of the members of that set has a probability greater than or equal to *k*. This latter claim is equivalent to saying that the rule

$$\text{If “B}ap \text{ \& B}aq\text{”} \in \text{U}a\text{, then “R}a(p \text{ \& } q)\text{”} \in \text{U}a$$

is acceptable. This rule is easily established by reference to (C.BR) and the fact that there is a closure condition for the set of sentences believed by *a*. Thus, even though there is no closure condition on *R*, the closure principle for belief places definite limitations on what *a* can consistently believe even if he restricts his beliefs to propositions that are reasonable. In the lottery case, *a* is obligated to believe *q* (i.e., that at least one ticket will be drawn) and is permitted to believe any proposition or set of propositions that satisfies the above provisos. Of course, *a* is also permitted to refrain from believing any of these propositions, with the exception of *q*.

The fact that *a* is permitted to believe propositions other than *q*, in the lottery case, places this theory in conflict with one prevalent attitude toward such cases. According to this attitude, the only rational thing to do in cases like the lottery case is suspend belief; that is, to believe *q* but refrain from believing anything else. Now to suspend belief is to refuse to believe anything beyond the evidence available to you; the attitude lying behind this refusal seems to be that the only propositions it is rational to believe are those that are acceptable as evidence. One who advocates suspension of belief in the lottery case might, in other words, appeal to the principle

(10) It is rational for *a* to believe that *p* if and only if *p* is acceptable as evidence for *a*.

And, given the schematic account of the concept Evident in previous pages, the following seems plausible:

(11) *p* is acceptable as evidence to *a* if and only if *p* is evident for *a*.

The obvious and immediate result of (10) and (11) is that propositions which are *merely* reasonable are not acceptable as evidence and thus cannot be rationally believed. This result has the effect of collapsing the distinction between obligatory and permissible belief, for the only permissible beliefs would be those that are obligatory. In the language of my theory, (10) and (11) would support the C-rule

(12) "Bap" $\in$ Ua if and only if "Eap" $\in$ Ua.

But (12), if added to the extant list of rules would yield

(13) "Eap" $\in$ Ua if and only if "Rap" $\in$ Ua.

The resulting general theory of rational belief would differ significantly from the one that I have advanced.

The problem of when one ought to suspend belief is, in general, a crucial problem for theories of rational belief. The conflict between the theory that I have suggested and that embodied in the rules (12) and (13) is representative of two primary attitudes one can take toward this problem. Suspension of belief is generally associated with 'neutrality' of evidence. That is, if one's evidence is 'neutral' with respect to a set of competing hypotheses, then one ought to suspend belief. But a man's evidence can be neutral in at least two ways; namely, (a) if the evidence fails to indicate (i.e., to render evident) any of the competing hypotheses, and (b) if the evidence fails to sustain (i.e., render reasonable) any of the competing hypotheses. If a man's evidence is neutral in sense (b), then it is neutral in sense (a), but not vice versa. Those who support a principle such as (10) will probably also construe neutrality of evidence in sense (a), while the theory that I have represented construes it in sense (b).

The simplest cases are those in which the competing hypotheses are p and its denial "$\sim p$". According to my theory, neutrality of the evidence available to a would be represented by

"$\sim$ Rap & $\sim$ R$a \sim p$" $\in$ Ua,

while the conflicting theory would represent neutrality by

"$\sim$ Eap & $\sim$ E$a \sim p$" $\in$ Ua.

I am inclined to say, however, that most cases in which neither p nor "$\sim p$" is evident are better described as cases in which the evidence is *inconclusive*. In general, a man's evidence might fail to render any of a set of competing hypotheses evident for him even though it provides strong support for one or more of those hypotheses. It seems inappropriate to call such evidence 'neutral'; more appropriately, we should say that such evidence fails to place one under any definite doxastic commitments, and from that point of view is inconclusive. Such situations should be distinguished, in my opinion, from those in which the evidence is genuinely neutral; that is, where none of the competing hypotheses is significantly supported by the evidence.

These reflections lead to one sort of objection that is likely to be brought against the results of my theory; namely, that they are overly libertarian. While the theory does allow for a considerable degree of doxastic freedom, I do not feel that this objection is telling. In the first place, even though suspension of belief is not obligatory in the lottery case, and similar cases, one who is inclined to order his beliefs in accordance with a strict set of principles such as (12) and (13) is perfectly free to do so; and one who is not so inclined is still required to keep his beliefs, as a set, within the boundaries of the reasonable. In the second place, suspension of belief *is* required by this theory in some cases, namely those in which *a*'s evidence is neutral in sense (b). For example, if one is considering whether a well balanced coin will come up heads or tails on the next toss, then suspension of belief is obligatory. The competitive hypotheses "heads will come up" and "tails will come up" each have a probability of .5 (or very close to that) and thus neither of them is reasonable.[13] In the third place, and most importantly, even though principle (10) is rejected by my theory, principle (11) is accepted. Thus, to say that *a* can (is permitted to) believe propositions other than *q* in the lottery case does *not* mean that he is permitted to accept them as evidence. The rule, or criterion, that one uses to determine which propositions are evident can perhaps be used as a 'detachment' rule; but the criterion for reasonable belief is not in general applicable as a rule of detachment. The corpus of propositions that one can rationally believe is not identical with one's evidential base, although the latter is included in the former.

Intuitions regarding suspension of belief are likely to be mixed, perhaps especially with regard to cases like the lottery paradox. It is tempting to assume that in a given cognitive situation there will be a fixed set of propositions, namely those that are acceptable as evidence, such that one is rational if and only if he believes all and only the members of that set. On the other hand, there is intuitive appeal to the suggestion that in many cognitive situations one is free to choose among the members of a set of competing hypotheses, provided one's choice is limited to propositions that are sustained by the evidence. The theory that I have constructed is designed to fix the limits of our doxastic freedom, in accordance with the second of these approaches.

IV. CONCLUSION

The scope of this discussion has been limited in many ways. For one thing, I have restricted my attention to the qualitative concepts Evident, Reasonable, Unreasonable, and Believed. Some philosophers feel that these concepts should be replaced by comparative concepts such as 'more reasonable than', or by quantitative concepts such as 'reasonable to degree *k*' and 'rational degree of confidence'. It is by no means clear, in my opinion, that we need to make a final choice among these various kinds of concepts; each of them appears to have its proper function in problems that arise regarding rational belief. A completely adequate theory of rational belief ought to tell us not only what we are obligated to believe, but also what degrees of confidence we are obligated (or permitted, or forbidden) to have in the things that we believe. Such a theory might constitute an addition to, rather than a replacement for, the one that I have developed here.

I have also limited my discussion to the consistency of rational belief at a given time; in reality, a man's beliefs are constantly changing – and so is his evidence. I earlier characterized the sets called 'epistemic universes' as cross-sections of an epistemic 'biography'; the rules that were provided do not apply to the biography itself, nor to the (rational) movement from one stage of a biography to the later ones.

Finally, I have avoided altogether the problems that arise when quantifiers are mixed in with our epistemic 'modalities'. A good deal of work has been done on the rules required for quantifiers in the logic of knowledge and belief; there are intriguing problems involved in the mixing of quantifiers with 'E*a*', 'R*a*', and 'U*a*' as well. For example, consider the expressions

(viii) "(∃x) E*a* (Fx)".
(ix) "E*a* (∃x) (Fx)".

What are the implications that hold between these expressions? In order to make our theory adequate we should account for such problems.

These are some of the primary problems that I have not discussed in this paper; they are interesting problems, and they deserve discussion, for the answers to them are not at all clear. It is hoped that the approach to the theory of rational belief developed here, in terms of the concept of epistemic consistency, will prove fruitful in the search for these answers.

APPENDIX

A. *A Formal Characterization of Epistemic Consistency*

Given the definitions of doxastic and evidential alternatives in the text, we define the relation of *accessibility* as follows:

> An alternative U*a** is accessible from an epistemic universe U*a* (with respect to *a*) if and only if U*a** can be reached in a finite number of steps each of which goes to an alternative of U*a* (with respect to *a*), in accordance with the C-rules. (Cf. Hintikka, *Knowledge and Belief*, p. 46.)

We then define epistemic consistency:

> An epistemic universe U*a* is epistemically consistent if and only if the set consisting of U*a* + the alternatives accessible from U*a* (with respect to *a*) satisfies the C-rules.

An epistemic universe is inconsistent if and only if it is not consistent.

B. *Theorems*

In the text, the following C-rules were adopted:

(C.cons) If *p*1 ∈ U*a* and *p*2 ∈ U*a* and ... and *pn* ∈ U*a*, then if the conjunction of *q* with *p*1 and *p*2 and ... and *pn* entails a contradiction, it is not the case that *q* ∈ U*a*.

(C.ded) If *p*1 ∈ U*a* and *p*2 ∈ U*a* and ... and *pn* ∈ U*a*, then if the conjunction of *p*1 and *p*2 and ... and *pn* entails *q*, then *q* ∈ U*a*.

(C.C*) If "C*ap*" ∈ U*a*, then there is at least one doxastic alternative U*a** to U*a* such that *p* ∈ U*a**.

(C.BB*) If "B*ap*" ∈ U*a*, then for every doxastic alternative U*a** to U*a*, "B*ap*" ∈ U*a**.

(C.B*) If "B*ap*" ∈ U*a*, then for every doxastic alternative U*a** to U*a*, *p* ∈ U*a**.

(C.∼C) If "∼ C*ap*" ∈ U*a*, then "B*a* ∼ *p*" ∈ U*a*.

(C.L*) If "L*ap*" ∈ U*a*, then there is at least one evidential alternative U*a** to U*a* such that *p* ∈ U*a**.

(C.EE*) If "E*ap*" ∈ U*a*, then for every evidential alternative U*a** to U*a*, "E*ap*" ∈ U*a**.

(C.E*) If "E*ap*" ∈ U*a*, then for every evidential alternative U*a** to U*a*, *p* ∈ U*a**.

(C.∼L) If "∼ L*ap*" ∈ U*a*, then "E*a* ∼ *p*" ∈ U*a*.

(C.E) If "E*ap*" ∈ U*a*, then "∼ E*a* ∼ *p*" ∈ U*a*.

(C.Rded) If "R*ap*" ∈ U*a* and *p* entails *q*, then "R*aq*" ∈ U*a*.

(C.E∼R) If "E*ap*" ∈ U*a*, then "∼ R*a* ∼ *p*" ∈ U*a*.

(C.U∼E) If "U*ap*" ∈ U*a*, then "∼ E*ap*" ∈ U*a*.

(C.R∼U) "R*ap*" ∈ U*a* if and only if "∼ U*ap*" ∈ U*a*.

(C.EB) If "E*ap*" ∈ U*a*, then "B*ap*" ∈ U*a*.

(C.BR) If "B*ap*" ∈ U*a*, then "R*ap*" ∈ U*a*.

[These rules are not independent; (C.E) and (C.U∼E) can be derived easily from the others.]

Given these, we can derive the following theorems (among others); ('∼ [...]' will mean '...is not a theorem'):

(T1) If "E*ap*" ∈ U*a* and *p* entails *q*, then "E*aq*" ∈ U*a*.

PROOF:

(1) "E*ap*" ∈ U*a* and *p* entails *q*. (assume)

(2) "∼ E*aq*" ∈ U*a*. (counterassumption)

(3) "L*a* ∼ *q*" ∈ U*a*. ((2), (C.∼L))

(4) "∼ *q*" ∈ U*a** where U*a** is an evidential alternative to U*a* (with respect to *a*). ((3), (C.L*))

(5) *p* ∈ U*a**. ((1), (C.E*))

(6) *q* ∈ U*a**. ((5), (1), (C.ded))

But (4) and (6) violate (C.cons).

(T2) If "E*ap*1 & E*ap*2 & ⋯ & E*ap*n" ∈ U*a*, then "E*a*(*p*1 & *p*2 & ⋯ & *p*n)" ∈ U*a*.

PROOF: This is an expanded version of the closure principle proven in Section III.

(T3) If "B*ap*" ∈ U*a* and *p* entails *q*, then "B*aq*" ∈ U*a*.

PROOF: Same as (T1) substituting "Cap" for "Lap" and letting Ua* be a doxastic alternative to Ua.

(T4) If "Bap1 & Bap2 &··· & Bapn" ∈ Ua, then "Ba(p1 & p2 &··· & pn)" ∈ Ua.

PROOF:

(1) "Bap1 & Bap2 &··· & Bapn" ∈ Ua. (assume)
(2) "∼ Ba(p1 & p2 &··· & pn)" ∈ Ua. (counterassumption)
(3) "Ca ∼ (p1 & p2 &··· & pn)" ∈ Ua. ((2), (C.∼B))
(4) "∼ (p1 & p2 &··· & pn)" ∈ Ua*, where Ua* is a doxastic alternative to Ua. ((3), (C.C*))
(5) "p1 & p2 &··· & pn" ∈ Ua*. ((1), (C.B*), (C.ded))

But (4) and (5) violate (C.cons).

(T5) ∼ [If "Rap1 & Rap2 &··· & Rapn" ∈ Ua, then "Ra(p1 & p2 &··· & pn)" ∈ Ua.]

PROOF: This is an expanded version of the proof given in Section III.

(T6) If "Eap1 & Eap2 &··· & Eapn" ∈ Ua, and if the conjunction of q with p1 and p2 and ... and pn entails a contradiction, then "Ea ∼ q" ∈ Ua.

PROOF: This follows from (C.∼L), given our interpretation of 'Lap'; it also follows via (T2) and (T1).

(T7) Analogue to (T6) for "Bap".

(T8) ∼ [If "Rap1 & Rap2 &··· & Rapn" ∈ Ua and the conjunction of q with p1 and p2 and ... and pn entails a contradiction, then "Ra ∼ q" ∈ Ua]

PROOF: Assume (1) that n=2, (2) that p1, p2, and q are pairwise consistent but conjoined are inconsistent, (3) that Ua contains all and only the members listed below and (4) that Ua is both a doxastic and an evidential alternative to itself and (5) there are no other alternatives to Ua:

Ua = {"Rap1 & Rap2", "∼ Ra ∼ q", "Rap1", "Rap2", "∼ Uap1", "∼ Uap2", "Ua ∼ q", "∼ Ba ∼ q", "Caq", "∼ Ea ∼ q", "Laq", q}.

Thus {"Rap1 & Rap2 &··· & Rapn", "∼ Ra ∼ q"} is consistently imbeddable in some Ua + accessible alternatives even though p1, p2, ..., pn and q conjoined entail a contradiction.

[As an example of Ua above, let $p1 = p$, $p2 =$ "$p \supset r$", and $q =$ "$\sim r$".]

(T9) If "Lap" $\in$ Ua and p entails q, then "Laq" $\in$ Ua.

PROOF:

(1) "Lap" $\in$ Ua and p entails q. (assume)

(2) "$\sim$ Laq" $\in$ Ua. (counterassumption)

(3) "E$a \sim q$" $\in$ Ua. ((2), (C.$\sim$L))

(4) $p \in$ Ua*, where Ua* is an evidential alternative to Ua. ((1), (C.L*))

(5) $q \in$ Ua*. ((4), (C.ded))

(6) "$\sim q$" $\in$ Ua*. ((3), (C.E*))

But (5) and (6) violate (C.cons).

(T10) If "Cap" $\in$ Ua and p entails q, then "Caq" $\in$ Ua.

PROOF: Analogous to the proof for (T9).

(T11) If "Eap $\vee$ aEq" $\in$ Ua, then "E$a(p \vee q)$" $\in$ Ua.

PROOF:

(1) "Eap $\vee$ Eaq" $\in$ Ua. (assume)

(2) "$\sim$ E$a(p \vee q)$" $\in$ Ua. (counterassumption)

(3) "L$a \sim (p \vee q)$" $\in$ Ua. ((2), (C.$\sim$E))

(4) "L$a(\sim p \;\&\sim q)$" $\in$ Ua. ((3), (T9))

(5) "$\sim$ L$a(\sim p \;\&\sim q)$" $\in$ Ua. (from 1 and (C.ded), given our reading of 'Lap': "Eap $\vee$ Eaq" entails "it is not compatible with everything that is evident for a that $\sim p \;\&\sim q$"; this could easily be established formally via a disjunctive syllogism, but it is quite obvious in any event.)

But (4) and (5) violate (C.cons).

(T12) Analogue to (T11) for "Bap".

PROOF: Similar to that for (T11).

(T13) $\sim$ [If "E$a(p \vee q)$" $\in$ Ua, then "Eap $\vee$ Eaq" $\in$ Ua.]

PROOF: We need to show that {"E$a(p \vee q)$", "$\sim$ (Eap $\vee$ Eaq)"} is consistently imbeddable in some Ua plus alternatives:

Ua = {"E$a(p \vee q)$", "$\sim$ (Eap $\vee$ Eaq)", "R$a(p \vee q)$", "B$a(p \vee q)$", "$\sim$E$a\sim(p \vee q)$", "$\sim$R$a\sim(p \vee q)$", "$\sim$ U$a(p \vee q)$", "$\sim$ Eap & $\sim$ Eaq", "L$a \sim p$", "L$a \sim q$", + whatever is required by (C.ded) and (C.Rded)}

Ua^* = {"$Ea(p \vee q)$", "$p \vee q$", "$\sim p$", "$Ba(p \vee q)$", + whatever is required by (C.ded)}

Ua^{**} = {"$Ea(p \vee q)$", "$p \vee q$", "$\sim q$", + whatever is required by (C.ded)}

(T14) Analogue to (T13) for "B*ap*".

(T15) Analogue to (T13) for "R*ap*".

(T16) If "B*ap*" ∈ U*a*, then "B*a*B*ap*" ∈ U*a*.

PROOF:

(1) "B*ap*" ∈ U*a*. (assume)

(2) "~ B*a*B*ap*" ∈ U*a*. (counterassumption)

(3) "C*a* ~ B*ap*" ∈ U*a*. ((2), (C. ~ B))

(4) "~ B*ap*" ∈ U*a**, where U*a** is a doxastic alternative to U*a*. ((3), (C.C*))

(5) "B*ap*" ∈ U*a**. ((1), (C.BB*))

But (4) and (5) violate (C.cons).

(T17) If "E*ap*" ∈ U*a*, then "E*a*E*ap*" ∈ U*a*.

(T18) If "E*ap*" ∈ U*a*, then "B*a*E*ap*" ∈ U*a*.

(T19) ~ [If "B*ap*" ∈ U*a*, then "E*a*B*ap*" ∈ U*a*.]

(T20) If "U*ap* & U*aq*" ∈ U*a*, then "U*a*(*p* & *q*)" ∈ U*a*.

(T21) If "U*a*(*p* ∨ *q*)" ∈ U*a*, then "U*ap* ∨ U*aq*" ∈ U*a*.

(T22) ~[If "U*a*(*p* & *q*)" ∈ U*a*, then "U*ap* & U*aq*" ∈ U*a*.]

The proofs for (T17)–(T22) have been omitted, since they are not difficult and resemble in structure the proofs given for other theorems.

University of Pennsylvania

BIBLIOGRAPHY

[1] Roderick Chisholm, 'Epistemic Statements and the Ethics of Belief', *Philosophy and Phenomenological Research* **16** (1956).

[2] Roderick Chisholm, 'The Logic of Knowing' (A Review of Hintikka's *Knowledge and Belief*), *The Journal of Philosophy* **60** (1963) 773–95.

[3] Roderick Chisholm, *Perceiving: A Philosophical Study*, Ithaca, N.Y., 1957.

[4] Roderick Chisholm, *Theory of Knowledge*, Englewood Cliffs, N.J., 1966.

[5] W. K. Clifford, *Lectures and Essays*, London 1886.

[6] Roderick Firth, 'Chisholm and the Ethics of Belief', *Philosophical Review* **68** (1959) 493–506.

[7] Herbert Heidelberger, 'Chisholm's Epistemic Principles', *Nous* **3** (1969) 73–82.

[8] Herbert Heidelberger, 'Knowledge, Certainty and Probability', *Inquiry* **6** (1963) 242–9.

[9] Carl Hempel, 'Deductive-Nomological versus Statistical Explanation', in *Minnesota Studies in the Philosophy of Science*, Vol. 3 (ed. by H. Feigl and G. Maxwell), Minneapolis 1962, pp. 98–169.
[10] Carl Hempel, 'Inductive Inconsistencies', *Synthese* **12** (1960) 439–69.
[11] Jaakko Hintikka, 'Form and Content in Quantification Theory', *Acta Philosophica Fennica* **8** (1955) 7–55.
[12] Jaakko Hintikka, *Knowledge and Belief*, Ithaca, N.Y., 1962.
[13] Jaakko Hintikka, 'A New Approach to Sentential Logic', *Societas Scientiarum Fennica, Commentationes Physico-Mathematica* **17** (1953), No. 2.
[14] Jaakko Hintikka, 'Notes on the Quantification Theory', *Societas Scientiarum Fennica, Commentationes Physico-Mathematica* **17** (1955), No. 12.
[15] Jaakko Hintikka and Risto Hilpinen, 'Knowledge, Acceptance, and Inductive Logic', in *Aspects of Inductive Logic* (ed. by J. Hintikka and P. Suppes), Amsterdam 1966, pp. 1–20.
[16] Henry E. Kyburg, Jr., 'A Further Note on Rationality and Consistency', *The Journal of Philosophy* **60** (1963) 463–5.
[17] Henry E. Kyburg, Jr., *Probability and the Logic of Rational Belief*, Middletown 1961.
[18] K. Lehrer, R. Roelofs, and M. Swain, 'Reason and Evidence: An Unsolved Problem', *Ratio* **9** (1967) 38–48.
[19] Leonard Linsky, 'On Interpreting Doxastic Logic', *The Journal of Philosophy* **65** (1968) 500–2.
[20] W. V. O. Quine, *Mathematical Logic*, Cambridge, Mass., 1951.
[21] Frederic Schick, 'Consistency', *The Philosophical Review* **75** (1966) 467–95.
[22] Frederick Schick, 'Consistency and Rationality', *The Journal of Philosophy* **60** (1963) 5–19.
[23] Marshall Swain, 'Schick on Consistency', *Philosophical Review* (forthcoming).

REFERENCES

* Much of the work on this paper was supported by University of Pennsylvania Faculty Research Grants, for which I am very grateful.

1 Throughout this paper double quote marks will be used in the way Quine [20] uses corner quotes; thus, "*a* believes that *p*" does not refer to the expression that occurs between the quote marks, but to the statement that results if we replace '*a*' with *a* and '*p*' with *p*. This notational device is borrowed from Hintikka (see note 4).

2 Frederic Schick [21], p. 469.

3 Some philosophers have suggested that these epistemic concepts *are* moral concepts. For a classic representative of this view see W. K. Clifford [5]. A more precise contemporary statement of this view can be found in Roderick Chisholm [3]. For an effective criticism of this view, see Roderick Firth [6]. On the basis of Firth's criticism's, Chisholm appears to have abandoned this view; see [4].

4 Hintikka's system is expounded at length in [12], and in a number of articles referred to in that book; see especially [11], [13], [14].

5 I am particularly concerned to avoid the sort of criticism raised by Leonard Linsky in [19]. Linsky's argument proves that there is a serious divergence between Hintikka's formal and informal characterizations of defensibility. For example, "B*a*B*b*(p & not-*p*)" is provably indefensible in Hintikka's system, but does not satisfy his informal charac-

terization of defensibility. Briefly, I avoid this problem by relativizing epistemic consistency to a given person *a* (more precisely, to the epistemic universe U*a*) and revising the formal characterization of epistemic consistency. This formal characterization is given at the beginning of the appendix to this paper.

6 See H. E. Kyburg, Jr., 'Conjunctivitis', in this volume. Kyburg distinguishes 'weak' from 'strong' versions of several principles governing logical properties. In his discussion, these principles apply to sets of *reasonably accepted* sentences; I have taken the liberty of adapting his terminology to the sets I am calling 'epistemic universes'.

7 The concept of a doxastic alternative is borrowed from Hintikka [12], p. 49.

8 I am grateful to my colleague Brian Chellas for pointing out the potential of (C.dens).

9 Henry E. Kyburg, Jr. [17], pp. 197–8.

10 The term 'justified' that occurs in these principles is one that we have not dealt with in this paper. It is a term often used in discussions of rational belief, however. My argument is based partly on an ambiguity in this term and I have deliberately used it in order to keep my presentation of the lottery paradox as innocent as possible.

11 This suggestion is made, for example, by Schick, *op. cit.*

12 The relationship of accessibility is defined in the Appendix to this paper.

13 It is worth pointing out again that my claim here depends on the schematic explication of Reasonable given in Section I.B. If I were to change the definition of 'sustains' so that k can be *equal to* or greater than .5, then it seems doubtful that suspension of belief would ever be obligatory.

HENRY E. KYBURG, JR.

CONJUNCTIVITIS*

I

Consider a set S of statements that may be taken to represent an idealized body of scientific knowledge. Let s_1 and s_2 be members of S. Should we regard the conjunction of s_1 and s_2, also as a member of S? It is tempting to answer in the affirmative, and a number of writers, whose systems we shall consider below, have indeed answered this way. An affirmative answer is conjunctivitis, which may be expressed by the following principle:

The Conjunction Principle: If S is a body of reasonably accepted statements, and s_1 belongs to S and s_2 belongs to S, then the conjunction of s_1 and s_2 belongs to S.

This principle is clearly equivalent to the following principle.

The Conjunctive Closure Principle: If S is a body of reasonably accepted statements, then the conjunction of any finite number of members of S also belongs to S.

Already the intuitive plausibility of the conjunction principle begins to fade; while it seems reasonable enough to want to accept the conjunction of two relatively elementary statements that are individually acceptable, it seems quite unreasonable to accept all the enormously long conjunctions of elements in S. But the reasonableness or unreasonableness of the principle will depend, of course, on what other principles one also accepts.

One principle which is, so far as I know, universally accepted[1], is the principle that anything entailed by a member of S should also be a member of S. If it is reasonable to accept s_1, and s_1 entails s_2, then it is reasonable to accept s_2. I shall call this the weak deduction principle.

The Weak Deduction Principle: If S is a body of reasonably accepted statements, and s_1 belongs to S, and $s_1 \supset s_2$ is a theorem of our underlying logic, then s_2 belongs to S.

Another principle which I am sure is universally accepted is that the set of reasonably accepted statements S should contain no contradictions.

Swain (ed.), Induction, Acceptance, and Rational Belief.

Whether or not it is psychologically possible to believe a contradiction (with practice perhaps it is), we do not want to regard it as rational. This I shall call the weak consistency principle.

The Weak Consistency Principle: If S is a body of reasonably accepted statements, then there is no member of S that entails every statement of the language.

In 1961 I offered an argument from the weak consistency principle and the weak deduction principle to the denial of the conjunction principle. This argument has come to be called the lottery paradox, and has engendered a number of principles designed to restrict the contents of S in such a way that all three of the principles mentioned so far hold.

The argument is this: Consider a fair lottery with a million tickets. Consider the hypothesis, 'ticket number 7 will not win'. Since this is, by hypothesis, a fair lottery, there is only one chance in a million that this hypothesis is false. Surely, I argued, this is reason enough to accept the hypothesis. But a similar argument would provide reason to accept the hypothesis that ticket i will not win, no matter what ticket number i may be. By the conjunction principle, we obtain from 'ticket 1 will not win' and 'ticket 2 will not win', the statement 'neither ticket 1 nor ticket 2 will win'; from the last statement, together with the statement 'ticket 3 will not win', by the conjunction principle, 'neither ticket 1, nor ticket 2, nor ticket 3 will win'; and so on, until we arrive at the reasonable acceptance of a long conjunction which can be briefly expressed as: 'For all i, if i is a number between one and a million, inclusive, ticket i will not win'. But we may also suppose that S contains the statement that the lottery is fair; and this statement entails the statement: 'For some i, i is a number between one and a million inclusive, and ticket i will win'. By the weak deduction principle we must therefore include this latter statement in S. By the conjunction principle we must therefore include the conjunction of the universally quantified statement and the existentially quantified statement in S. But this conjunction is an explicit contradiction from which any statement will follow in violation of the weak consistency principle. I concluded that it was worth while to hang onto the weak deduction principle and the weak consistency principle, and therefore that the conjunction principle should be abandoned.

Quite a number of people, finding the conjunction principle more plausible than I do, have attempted to spike this argument here or there.

One of the earliest attempts was made by Salmon [14], who suggested that one ought not to accept particular statements (such as 'Ticket 7 will not win the lottery'), but restrict one's acceptances to general statistical or universal generalizations. Since a number of writers have followed Salmon in this ploy, it is worth stating a statistical version of the same argument.[2]

Consider a finite population P of entities, each of which either has or lacks a certain quality Q. We draw a random sample (in any sense of 'random' you choose) of n of the P's. A certain proportion of the members of the sample, f, have the property Q; we know that in the parent population P some unknown proportion p have the property Q. Now consider hypotheses of the form 'p lies in the interval i'; for example: 'p lies in the interval $(f-.1, f+.1)$', 'p lies in the interval $(f-.0001, f+.675)$', etc. There are a number of principles of inference that one might adopt for arriving at acceptable statistical statements of this form. I shall consider two, though what I say will apply to other principles as well. Let us call them the Bayesian Acceptance Principle and the Classical Acceptance Principle.[3] In accordance with the Bayesian Principle, we shall accept a statistical hypothesis if its posterior probability is greater than $1-\varepsilon$, i.e., if the probability of its negation is less than ε. According to the Classical Principle, we will accept a hypothesis provided the probability of rejecting it by mistake is less than ε. Since the argument is slightly different in the two cases, I shall treat them separately.

Bayesian case: There are any number of intervals i such that the hypothesis '$p \in i$' is acceptable, under the assumption that the prior distribution of p is continuous between 0 and 1. Let i_c be the intersection of all these intervals. By the conjunction principle, '$p \in i_c$' is acceptable. Again under the assumption of continuity, it is possible to divide any interval i and in particular i_c into a finite number of subintervals $i_1, i_2, \ldots, i_m$, such that the posterior probability of '$p \in i_k$' is less than ε, for all k, $1 \leqslant k \leqslant m$. But this is just to say that the posterior probability of '$\sim p \in i_k$' is greater than $1-\varepsilon$ for all k, $1 \leqslant k \leqslant m$, and thus that the hypothesis '$\sim p \in i_k$' is acceptable for all k, $1 \leqslant k \leqslant m$. The conjunction principle then entails that '$\sim p \in i_c$' is acceptable, in virtue of the fact that $i_c = \bigcup i_k$. Thus by the conjunction principle we have '$p \in i_c \,\&\, \sim p \in i_c$' in our body of acceptable statements in violation of the weak consistency principle.

Classical case: Again there are any number of intervals i such that the probability that we will falsely reject the hypothesis '$p \in i$' when we observe f is less than ε. To be more precise (and more classical), to each of these intervals i_j there will correspond an interval E_j, such that if we reject the hypothesis '$p \in i_j$' if and only if the observed frequency f falls outside the interval E_j, then we will falsely reject the hypothesis no more than ε of the time. Of those hypotheses '$p \in i_j$' such that in point of fact f falls in the corresponding interval E_j, we say that they 'are not rejected at the ε level of significance'. In particular, to each hypothesis of the form '$p \in (a, b)$' there will correspond a test interval $(a-d, b+e)$ (it will always include the *closed* interval $[a, b]$), such that if we reject the hypothesis if and only if we observe a value of f not falling in the test interval, we shall falsely reject it less than ε of the time. Consider two hypotheses '$p \in (a, f)$' and '$p \in (f, b)$'. At any level of significance, the value f of the observed frequency will fall within the test interval corresponding to each of these hypotheses.[4] Thus at any level of significance we will accept both the hypothesis '$p \in (a, f)$' and the hypothesis '$p \in (f, b)$'. By the conjunction principle, we must then accept their conjunction. But there is no number that belongs both to (f, b) and (a, f); again we find a violation of the weak consistency principle.

II

Among those who have adopted principles of acceptance satisfying the principle of conjunction, the weak principle of consistency, and the weak principle of deduction, are included Hempel [3], Hintikka [7], Lehrer [12], and Levi [13]. In the ensuing sections I shall consider the principles proposed by each of these writers in turn, both from the viewpoint of strength and from the viewpoint of plausibility. We begin with Hempel. In his well known paper, 'Deductive-Nomological vs. Statistical Explanation', Hempel lays down certain "necessary conditions of rationality in the formation of beliefs". I shall discuss them one by one.

(CR1) Any logical consequence of a set of accepted statements is likewise an accepted statement; or, K contains all logical consequences of any of its subclasses.

An obviously equivalent formulation is the following:

The Principle of Deductive Closure: The set S is closed under deduction.

It should be observed that this principle is not equivalent to the weak deduction principle. Hempel's criterion entails the weak deduction principle, but the converse does not hold. For example, the set: {'S', '$S \supset P$', '$S \vee \phi$', '$\sim P \supset \sim S$', '$S \& T$', 'T',...} of which 'P' is not a member satisfies the weak principle, but not Hempel's strong principle. However, the conjunction principle, together with the weak deduction principle, are equivalent to the principle of deductive closure:

THEOREM 1: Principle of deductive closure≡(principle of conjunction and weak deduction principle)

PROOF: If S satisfies the principle of deductive closure, then if s_1 and s_2 belong to S, so do all their consequences; among their consequences is the conjunction of s_1 and s_2. Thus the principle of conjunction. And if S satisfies the principle of deductive closure, then if s_1 belongs to S, so does any consequence of s_1. Thus the weak principle of deduction. Now suppose S satisfies both the principle of conjunction and the weak deduction principle; then suppose that $s_1, \ldots, s_n$ each belong to S, and that s follows from $s_1, \ldots, s_n$ as premises. By the principle of conjunction, the conjunction of $s_1, \ldots, s_n$ belongs to S; by the standard deduction theorem, plus the weak deduction principle it then follows that s belongs to S.

What is questionable about the strong principle of deductive closure is, I think, precisely the principle of conjunction. This may involve a matter of intuition: I simply don't believe that everything I believe on good grounds is true, and I think it would be irrational for me to believe that. Other people seem to think the opposite. I suspect that at root there is a confusion of quantifiers: of everything that I believe, it is correct to say that I believe it to be true; but it is not correct to say that I believe everything I believe to be true. In symbols:

(1) (x) (I believe $x \supset$ I believe x to be true)
(2) I believe (x) (I believe $x \supset x$ is true).

Statement (1) seems true, statement (2) false.

Hempel's second condition of rationality is a consistency condition

(CR2) The set K of accepted statements is logically consistent.

Although it might be possible to construe this as the weak principle of consistency above, it is more likely that Hempel has in mind a strong principle of consistency:

Strong Principle of Consistency: If S is a body of reasonably accepted statements, then there is no finite subset of S, $s_1, \ldots, s_n$, such that every statement of the language follows from $s_1, \ldots, s_n$ as premises.

Again conjunction plays an important role. In the presence of the conjunction principle, the strong principle of consistency and the weak principle of consistency are equivalent.

THEOREM 2: Conjunction principle $\supset$ [strong principle of consistency $\equiv$ weak principle of consistency].

Again one can question the plausibility of the strong principle. I probably cannot believe a contradiction, or act on one. But I can certainly believe, and even act on, each of a set of statements which, taken conjointly, is inconsistent. Indeed, when I lend my moral support to a lottery, without buying a ticket, this is one way to describe what is going on; though a more adequate and complete description would involve expectation.

Hempel's third criterion is, he claims, "simply a restatement of the requirement of total evidence" (p. 151).

(CR3) The inferential acceptance of any statement h into K is decided on by reference to the total system K.

This criterion, though it is not strictly relevant to the problem of conjunctivitis, is worth a comment or two. It is clearly in conflict with the principles of fallibilism and empiricism which underly much of what Hempel writes. Surely it is a consequence of these principles that even if a statement s becomes a part of our body of reasonable beliefs, we will, if the evidence begins to go the other way, be ready to reject it. But if we incorporate the statement h into the body of beliefs K, then, whatever else we add to that body of beliefs, its probability, relative to that body of beliefs, is going to be unity. Once accepted, no added evidence can ever render h improbable. The suggestion – which requires a great deal of development – is that we shall have to keep our evidential base separate from the body of reasonable beliefs erected on that base. There is also a new path to inconsistency opened by this principle. Let us suppose that there is a set P, 99% of which are Q_1; that 99% of the Q_1 are Q_2; ... that 99% of the Q_n are R. There is nothing inconsistent in supposing this, and that at the same time 99% of the P are not-R. But there is obviously something inconsistent about accepting the statement that an arbitrary P is not-R (since it is a P and that's all we know about it), and accepting the statement that an arbitrary P is R (since it is a P and that's all we know about it).

Hempel finally proposes a measure of epistemic utility (related to the content of a statement in terms of a logical measure function) according to which the principle of maximizing utility leads to the rule:

> Tentative rule for inductive acceptance: Accept or reject h, given K, according as $c(h, K) > 1/2$ or $c(h, K) < 1/2$; when $c(h, K) = 1/2$, h may be accepted, rejected, or left in suspense (p. 155).

It is clear that this rule is not only, as Hempel puts it, "too liberal", but leads directly, through the strong principle of deduction, to inconsistency, provided only that there are three hypotheses, jointly exhaustive, whose probabilities, relative to k are all less than a half.

III

A recent paper by Hilpinen and Hintikka develops an inductive acceptance rule that is demonstrably consistent, and satisfies a number of criteria similar to those discussed above. Their system uses Hintikka's 1965 system of inductive logic [4], in which non-zero degrees of confirmation come to be assigned to general sentences. This system of inductive logic is applicable in principle to all first order languages (without identity), but is developed in detail only for monadic languages. In point of fact, the system developed by Hilpinen and Hintikka satisfies the strong consistency principle, and the strong principle of deductive closure. Put in terms of the most finely articulated statements above: their system satisfies the weak deduction principle, the weak consistency principle, and the conjunction principle. (Since these three principles are independent, it seems best to refer to them separately.)

Since Hintikka's system is not as well known as it should be, a brief review of its features may be helpful here. Consider a language L, containing k primitive monadic predicates 'P_i'. By means of these predicates, one can characterize $K = 2^k$ kinds of individuals, corresponding to Carnap's 2^k Q-predicates. For the sake of simplicity we suppose that instantiation of each of the Q-predicates is logically possible, or in Carnap's terms, that the primitive predicates are logically independent. These Q-predicates are called by Hintikka 'attributive constituents'. A *constituent* consists of a specification, for each of the K Q-predicates or attributive constituents, of whether or not it is instantiated in the world. There are 2^K-1 possible different constituents (because it is logically false that *no* constituent be instantiated in a non-empty universe).

There are various ways of assigning a priori probabilities to the constituents. One might take the probability of a constituent to be proportional to $(w/K)^\alpha$, where α is an arbitrary constant, and w is the number of Q-predicates that are alleged to be instantiated by the constituent. For our purposes the simpler assignment of equal probabilities to each constituent (also worked out by Hintikka and Hilpinen) will suffice.

Let e be a sentence that asserts, for each member of a sample of n individuals, and for each of our primitive predicates 'P_i', either that that individual has the property P_i or that it has the property $\sim P_i$. Since we may order the Q-predicates in an arbitrary way there is no loss of generality in supposing that our sample of n individuals provides instantiation of the first c Q-predicates. The constituents that are consistent with our evidence all have the form:

$$(\exists x)\, Q_1(x)\ \&\ (\exists x)\, (Q_2(x))\ \&\cdots\&\ (\exists x)\, Q_c(x)\ \&\ (\exists x)\, Q_{i_1}(x)$$
$$\&\ (\exists x)\, Q_{i_2}(x)\ \&\cdots\&\ (\exists x)\, Q_{i_m}(x)\ \&\ (x)\, (Q_1(x) \vee Q_2(x)$$
$$\vee \cdots Q_c(x) \vee Q_{i_1}(x) \vee \cdots \vee Q_{i_m}(x)),$$

where $i_j < i_k$ whenever $j < k$, and $i_1 > c$.

Let C_c be the constituent that asserts that just those Q-predicates instantiated by our evidence are exemplified in the universe (i.e., $C_c =$ '$(\exists x)\, Q_1(x)\ \&\cdots\&\ (\exists x)\, Q_c(x)\ \&\ (x)\, (Q_1(x) \vee \cdots \vee Q_c(x))$', and let C_w be any other constituent consistent with our evidence. Then it is possible to show:

(1) $P(C_c, e) > P(C_w, e)$
(2) $\lim_{n\to\infty} P(C_c, e) = 1$
(3) $\lim_{n\to\infty} P(C_w, e) = 0.$

One further fact is important:

(4) Every consistent general sentence h of L_k can be transformed into a disjunction of constituents; thus $P(h, e) = \sum_i P(C_i, e)$, where the summation is extended over all those constituents in the disjunction equivalent to h.

It is possible to show ((2) and (3) give an intuitive justification) that for given $\varepsilon < 1/2$, we can calculate an integer n_0 such that if $n > n_0$, one and only one constituent will have a probability, relative to e, greater than $1-\varepsilon$. This constituent will of course be C_c.

We now adopt the following for our rule of acceptance: Accept a *general* statement h, given evidence e, if and only if: the probability of h is greater than $1-\varepsilon$, and more than n_0 objects have been examined. Formally (p. 11):

$$\text{(D.Ac)} \quad Ac(h, e) =_{\text{Df}} \begin{array}{l} \text{(i)}\ P(h, e) > 1 - \varepsilon, \text{ where } 0 < \varepsilon \leqslant 0.5 \\ \text{(ii)}\ n > n_0. \end{array}$$

In virtue of (4) a hypothesis h can have a probability greater than $1-\varepsilon$ only if the constituent C_c appears in its distributive normal form; thus every hypothesis that is acceptable must be consistent with C_c; and thus they must be jointly consistent. The strong consistency condition, and thus the weak consistency condition also, is satisfied.

The principle of conjunction is also satisfied. If h_i and h_j are acceptable, then C_c must occur in the distributive normal form of each of them, and thus C_c will occur in the distributive normal form of their conjunction. The probability of their conjunction will therefore be greater than $1-\varepsilon$, and their conjunction will therefore be acceptable.

The principle of deductive closure so far fails, however, even in its weak form. Let '$(x)A(x)$' be an acceptable generalization; one of its deductive consequences is '$A(a)$'. But there is no clause in (D.Ac) that will allow us to accept '$A(a)$'.

Deductive closure does hold, however, for *general* statements. Suppose that the general statements $h_1, \ldots, h_n$ are all acceptable, and that they entail a factual general statement h. Since C_c occurs in the distributive normal form of each h_i, C_c must also occur in the distributive normal form of h itself. Thus h must be acceptable, and, so far as general statements h (containing no individual constants) are concerned, the strong principle of deductive closure is satisfied.

THEOREM 3: If K is the set of all those statements h such that $Ac(h, e)$, then K satisfies the conjunction principle and the strong consistency principle, but neither the strong nor the weak deduction principle. The strong and weak deduction principles are satisfied if we restrict their range of application to completely general sentences.

The system so far provides us with no way of accepting singular statements of the form $M(a_i)$, where M is a molecular predicate equivalent to a disjunction of Q-predicates, and a_i an individual constant. Hintikka and Hilpinen show that in order to preserve consistency, our rule for the acceptance of singular predictions must be:

(D.Ac sing) A singular hypothesis $(A(a_i)$ is acceptable if and only if the generalization $(x)A(x)$ is acceptable (p. 18).

Let A_c be the disjunction of the c Q-predicates in C_c. '$(x)A_c(x)$' is acceptable, and the probability of the conjunction of any arbitrary number of statements of the form $A_c(a_i)$ is at least $1-\varepsilon$. Indeed:

$$\lim_{r\to\infty} P[A_c(a_1) \,\&\cdots\&\, A_c(a_r)] = P(C_c, e)$$

Let us look at consistency and deductive closure. Suppose the singular statements '$A_1(a_1)$', '$A_2(a_2)$',... '$A_n(a_n)$' are acceptable as being instances of acceptable universal generalizations, and that the compound statement S is deducible from them. Since '$(x)A_c(x)$' is the strongest universal generalization, '$A_1(a_1)$' is deducible from '$A_c(a_1)$', '$A_2(a_2)$' is deducible from '$A_c(a_2)$' etc., and the statement S is deducible from the conjunction '$A_c(a_1) \,\&\, A_c(a_2) \,\&\cdots\&\, A_c(a_n)$', therefore S is deducible from '$(x)A_c(x)$'. This ensures that the probability of S will be at least $1-\varepsilon$. But we have no grounds for either asserting or denying that S is acceptable; acceptability has been defined only for completely general statements, and for singular statements of the form $A(a)$. Deductive closure thus far fails. Acceptability in general, however, would presumably be defined in this way:

(Acc) A statement S is acceptable if and only if it belongs to every class K of statements, closed under deduction, containing e, such that each member of K that is completely general is acceptable by (D.Ac), and such that each statement acceptable by (D.Ac) is a member of K.

The separate principle (D.Ac sing) is deducible from this principle.

THEOREM 4: If K is the set of all statements h, such that h is acceptable by the principle (Acc), then K satisfies the conjunction principle, the strong consistency principle, and the strong deduction principle.

PROOF:

(a) Conjunction Principle: The conjunction of s_1 and s_2 belongs to every class K closed under deduction to which both s_1 and s_2 belong, and therefore to every class K closed under deduction and meeting further requirements as well to which both s_1 and s_2 belong.

(b) The Strong Consistency Principle: By hypothesis e is consistent, and consistent with the strongest acceptable generalization, '$(x)Ac(x)$', Other statements are obtained only by deduction, but deduction cannot introduce inconsistency.

(c) The Strong Deduction Principle: Since the classes K are closed under deduction the deduction principle is satisfied automatically.

The set of statements characterized as acceptable by this principle consists essentially of C_c and e, together with all of their deductive consequences. This set is essentially the set of deductive consequences of a single statement. We shall find this to be characteristic of those acceptance rules for which deductive closure holds, and which are demonstrably consistent.

There are certain shortcomings to the system of Hintikka and Hilpinen. In the first place, as it stands, it does not allow us to take account of the relative frequencies with which the Q-predicates are exemplified. In a similar system proposed by Hintikka [6] the probability that an individual a will have a certain molecular property M will depend on the relative frequencies with which Q-predicates have been observed to be exemplified; but in the system under discussion we can never accept $A(a)$ unless the universal generalization $(x)M(x)$ is acceptable. This is a general feature of these systems, and necessarily so. All of these systems accept the principle of conjunction, and it follows directly from the principle of conjunction that if an arbitrary predictive inference of the form '$A(a)$' is acceptable, then its universal generalization '$(x)A(x)$' is acceptable, at least when restricted in scope to the unobserved part of the universe of discourse. This should be stated formally.

THEOREM 5: If the principle of conjunction is accepted, and if for any arbitrary individual a among the unobserved individuals, the singular predictive inference '$A(a)$' is acceptable, then the universal generalization '$(x)A(x)$' is acceptable, when restricted in scope to unobserved individuals.

PROOF: Let the evidence e mention only k individuals, as failing to satisfy the predicate 'A'; let these individuals be $a_1 \ldots a_k$ (k may be 0). Then if the principle of conjunction is accepted, then if in general any singular predictive inference of the form '$A(a_i)$' ($i>k$) is acceptable, then every finite conjunction of statements of this form is acceptable, and the universal generalization '$(x)\left(\sim x \in \{a_1, a_2, \ldots, a_k\} \supset (A(x))\right)$' is acceptable in a finite language. For an infinite language, we need only note that '$A(a_{k+1})$' is acceptable, and that if the restricted universal generalization is acceptable for all individuals up to the nth, then by the principle of conjunction it is acceptable also for all individuals up to the $(n+1)$st. A plausible induction principle yields the conclusion. In systems such as Hintikka's without identity we cannot express a generalization that takes account of exceptions; but the same result follows in those systems for cases when $k=0$.

The problem of extending the system of Hintikka and Hilpinen to a full first order logic is very knotty. Some steps have been taken in this direction by Hintikka, who has defined constituents quite generally. Hintikka's approach to inductive logic is applicable in principle to all first order logics, though the definition of (Acc) would obviously have to be enormously complicated even to deal with a language containing two-place predicates. Tuomela [15] has begun the attempt to construct an inductive logic for monadic languages with identity. But in virtue of the fact that the essence of the acceptance rule is that we are directed to accept a single statement $(C_c \& e)$, together with all of its deductive consequences, any such system will be open to objections of the sort that will be applied below to proposals of Levi and Lehrer.

Finally, it should be observed that in order to do statistical inference, we need a general higher order logic. In particular, we need to be able to speak of the set of n-membered subsets of a set S, if not in general, at least for sets of fairly high order. We also need a language rich enough for measure theory. Such a language, of course, is enormously more powerful than anything hitherto considered by Hintikka, Hilpinen, or Tuomela. To be sure one must start somewhere. But it is difficult to see what principles, analoguous to those adopted for the monadic predicate calculus, could be used to avoid the statistical versions of the lottery paradox described above.

IV

The system of inductive acceptance described by Levi [13] has the overwhelming advantage of being applicable to very rich languages. It is thus the sort of system we can apply in the kinds of circumstances that we can actually find ourselves in. It can be applied, for example, to the problem of accepting statistical hypotheses on given evidence. Two new concepts must be explained before the rule can be stated. The most important and most novel concept is that of an *ultimate partition*. Levi argues that the inductive inference maker does not conduct his inquiries in a vacuum; he does so in a context determined in part by a felt need, a problem which he is seeking to solve. Thus before the inquiry starts, an investigator has an idea, which may be quite clear, or, unanalyzed, may be rather confused, as to what would constitute a relevant answer to his problem. These answers can be related to relatively atomic relevant answers, which constitute an ultimate partition, U in the following way. Following Levi, we suppose that there is a certain statement b which represents background information, not up for test in the inquiry, and a certain statement e, which represents the body of evidence of the inquiry. The ultimate partition U_e is a (usually finite) set of sentences in L (the language of the inquiry, which may be as rich as you please), such that each element of U_e is consistent with b and e, and such that the conjunction of b and e entails (i) that some member of U_e is true, and (ii) that at most one member of U_e is true, and (iii) every relevant answer is logically equivalent to the disjunction of zero or more members of U_e, where we understand the disjunction of zero members of U_e to be the conjunction (inconsistent with b and e) of all the members of U_e. The set of sentences M_e represents the canonical standardized list of relevant answers. For an ultimate partition containing n members, M_e is the set of 2^n statements, constructed by forming the disjunction (in alphabetical order) of m ($1 \leqslant m < n$) elements of U_e, and adding to that list S_e, the disjunction of all the members of U_e and C_e, the conjunction of all the members of U_e. The subscript 'e' reflects the evidence e. An initially ultimate partition would be designated by 'U', and its set of ultimate answers in canonical form by 'M'. Given some evidence e, the initially ultimate partition would be reduced by the deletion of any elements of U that were inconsistent with e. This produces a truncated ultimate partition U_e, and leads to a new (and correspondingly truncated) set of relevant answers M_e.

The other crucial concept is content. Content is defined relative to ultimate partitions. Each element of the initial ultimate partition is taken to have the same content, on the grounds that any difference in content would lead to a finer ultimate partition. To arrive at the conditional content of a hypothesis H, given certain evidence e, one merely applies the same principle to the truncated ultimate partition: one takes each element of U_e to have the same content. This leads to the conclusion that in general,

> $\text{cont}(H, e) = m/n$, where n is the number of elements in U_e, and m is the number of elements in U_e that are inconsistent with H (p. 70).

With these two concepts at hand, we can state Levi's Inductive Acceptance Rule (p. 86).

Rule A: (a) Accept b & e and all its deductive consequences.

(b) Reject all elements a_i of U_e, such that $p(a_i, e) < q$ cont $(\sim a_i, e)$ i.e., accept the disjunction of all unrejected elements of U_e as the strongest element in M_e accepted via induction from b & e.

(c) Conjoin the sentence accepted as strongest via induction according to (b) with the total evidence b & e and accept all deductive consequences.

(d) Do not accept (relative to b, e, U_e, the probability distribution, and q) any sentences other than these in your language.

The number q referred to in the acceptance rule may be construed as an index of boldness; it ranges from 0 to 1, and the larger it is, the less cautious will one be in accepting statements not entailed by the background knowledge and evidence b & e. The number q reflects the "relative importance of the two desiderata: truth and relief from agnosticism".

As it stands, rule A obviously leads to the acceptance of a set of statements satisfying the strong principle of deductive closure (indeed Levi takes the condition of deductive cogency, as he calls it, as a condition of adequacy), and thus also the conjunction principle; the strong consistency principle is satisfied because q must be less than or equal to 1, and the probability level at which elements of U_e are rejected must therefore be less than $1/n$, where n is the number of elements of U_e. Even if q is taken to be 1, and the number of elements n of U_e is taken to be 2, the inequality in (b) preserves us from inconsistency. The general principle is the same as that embodied in Hintikka's system already discussed: what is accepted is essentially a single (strongest) hypothesis H, together with its deductive consequences.

Unlike Hintikka's system, in Levi's system a high probability is not necessary for acceptance. Levi's system has a rule of rejection which is not purely probabilistic, but is dependent on content (as determined by the number of elements in the ultimate partition) and on the index of caution q. One may perfectly well end up accepting a proposition on the evidence whose probability relative to that evidence is less than 1/2, so long as it is more probable than any competing alternative. The canonical example is that in which there are three elements in U_e, each (therefore) having content 1/3, q is 1, and the probabilities of a_1, a_2 and a_3 are respectively 0.4, 0.3, and 0.3. We are directed to accept a_1 and all the deductive consequences of $a_1 \& b \& e$. Strong consistency is nevertheless preserved.

While Hintikka's system appears to be the prototype of a global approach to problems of belief and acceptance, and thus to be limited by the choice of a language, and open to the difficulties of attempting to develop a similar system for richer languages, Levi's system is frankly local and problem dependent. Another way of construing the relation is to say that for Hintikka, the language we use determines the ultimate partition, while for Levi the ultimate partition is determined both by a language and by a particular problem; or perhaps it could be put this way: the partition is determined by a particular problem together with the language in which we represent that problem to ourselves.

The fact that Levi's rule of acceptance is relativized to a given language L, background knowledge b, evidence e, and a probability distribution P, raises no eyebrows; we surely expect that what a plausible acceptance rule will dictate will depend on these factors. That what we accept should depend in some way on how cautious or bold we are being, as expressed by the number q, also seems reasonable enough. The important question concerns the relativization to an ultimate partition. Given the language, b, e, the probability distribution and q, it is clear that different ultimate partitions may lead to the acceptance of different sets of statements. It would be blatantly contrary to the whole pragmatic spirit of Levi's approach to suppose that there is some special, preferred, universal, ultimately ultimate partition. All ultimate partitions must be treated on a par, though at a given time, under given circumstances, we may not *consider* or *think about* more than one. But we could consider several. Levi asks rhetorically, "...why is it impossible for conditions K and K'

to prevail at the same time, such that a man believes a deductively consistent and closed system of sentences Γ based on K and simultaneously a deductively consistent and closed system of sentences Γ' based on K'? To be sure he will believe, and believe rationally, all sentences in the set $\Gamma \cup \Gamma'$, and this set may very well be neither consistent nor [deductively] closed. But ... why should this be objectionable?" (p. 94).

One answer to this rhetorical question might consist in quoting the arguments Levi aduces elsewhere in his book in favor of the Principle of Deductive Cogency. Whatever reason there is for demanding that the set Γ based on K should be deductively consistent and closed, are these not also reasons for demanding that $\Gamma \cup \Gamma'$ be deductively consistent and closed? But I rather agree with Levi that there is nothing at all objectionable about a man believing rationally all the sentences in the set $\Gamma \cup \Gamma'$, where this set is neither consistent nor deductively closed. Indeed, since for any hypothesis that has a high probability, we can construct an ultimate partition which will lead to the acceptance of that hypothesis, we are in essentially the same state, so far as statements with high probability are concerned, as we are in one of my rational corpora. Levi points out (p. 95) that his Rule A "... requires that this set [of rationally accepted sentences] must be divisible into subsets which are consistent and closed relative to the total evidence and the ultimate partitions detached at that time". But since we can detach any partition we want, this latter requirement is empty; and it is trivial to divide any set of sentences satisfying my conditions for a rational corpus into subsets that are deductively cogent. Indeed, it is trivial to do this for any purely probabilistic rule of acceptance.

THEOREM 6: If $S \in K$ if and only if $\text{Prob}(S, e) > r$, then K may be divided into subsets which are consistent and closed – i.e., for which the strong consistency principle and the strong deduction principle are satisfied.

PROOF: For each sentence S of K, let K_S consist of S, together with all of its deductive consequences. The K_S are the required division. If $T \in K_S$, then T is entailed by S. But if S entails T, the probability of T cannot be less than that of S (on *any* interpretation of probability) and so T belongs to K. Conversely if $S \in K$, then $S \in K_S$. So $K = \bigcup K_S$. The sets K_S are deductively closed; they satisfy the Conjunction Principle. It remains to show that they are consistent. An inconsistency can appear only if it is entailed by some sentence S of K. But the probability of an inconsistency (on any interpretation of probability) is 0, and if S entails it, the probability of S cannot be greater than 0; and so S cannot be a member of K on purely probabilistic grounds.

Let us examine the set of statements that might come to be accepted (by means of some ultimate partition or other) in Levi's system. Levi points out that this set of statements will not in general satisfy either the strong deduction principle or the strong consistency principle. It is interesting to observe that these sets of statements do satisfy the weak deduction principle and the weak consistency principle.

THEOREM 7: If Γ_ω is the set of all those statements S such that there is an ultimate partition U_e, relative to which, and b, e, and q, S comes to be accepted by Rule (A), then Γ_ω satisfies the weak deduction principle and the weak consistency principle.

PROOF: Every statement that comes to be accepted in this set of statements is accepted originally in relation to some ultimate partition. But Rule A demands that when S is accepted relative to some ultimate partition, all of the deductive consequences of $S \& b \& e$ should also be accepted; thus the weak deduction principle. As for consistency, if S is inconsistent in itself, it can never come to be accepted by Rule A, and thus cannot be a member of the set of accepted statements relative to any ultimate partition.

There is one further kind of consistency we might ask about. The pair of statements $(S, \sim S)$ is perfectly consistent in the weak sense – i.e., it contains no inconsistent statement among its elements – and yet seems rather flatly wrong, somehow. What I shall call the Principle of Pairwise Consistency stipulates that a body of accepted statements should not contain any such pairs of statements.

Principle of Pairwise Consistency: If K is a body of reasonably accepted statements, then for no statement S of the language is it the case that both S and the denial of S belong to K.

In a similar manner we may define for every n a principle of n-wise consistency:

Principle of n-wise Consistency: If K is a body of reasonably accepted statements, then for no set of statements $s_1, s_2, \ldots, s_{n-1}$ is it the case that each of $s_1, s_2, \ldots, s_{n-1}$, and $\sim(s_1 \& s_2, \ldots, s_{n-1})$ is a member of K.

We should first observe that any purely probabilitistic acceptance rule (with acceptance level greater than 1/2) satisfies the Principle of Pairwise Consistency, and any purely probabilistic acceptance rule with acceptance level greater than $n/(n+1)$ satisfies the Principle of n-wise Consistency.

THEOREM 8: If $S \in K$ if and only if $\text{Prob}(S, e) > r$ (where $r > 1/2$), then K satisfies the Principle of Pairwise Consistency. If $S \in K$ if and only if $\text{Prob}(S, e) > n/(n+1)$ then K satisfies the Principle of n-wise Consistency.

PROOF: For ordinary point probabilities, the probability of S is one minus the probability of $\sim S$, so both probabilities cannot be greater than 1/2. For my interval probabilities, it is possible to show that $\text{Prob}(S) = (p, q)$ if and only if $\text{Prob}(\sim S) = (1-q, 1-p)$; so the same argument goes through.

$$\text{Prob} \sim [S_1 \mathbin{\&} S_2 \mathbin{\&} \cdots \mathbin{\&} S_{n-1}] = \text{Prob}[\sim S_1 \vee \sim S_2 \vee \cdots \vee \sim S_{n-1}]$$

$$\text{Prob}[\sim S_1 \vee \sim S_2 \vee \cdots \vee \sim S_{n-1}] \leqslant \text{Prob}(\sim S_1) + \text{Prob}[\sim S_2 \vee \sim S_2 \vee \cdots \sim S_{n-1}]$$

$$\text{Prob}(\sim S_1) + \text{Prob}(\sim S_2) + \cdots + \text{Prob}(\sim S_{n-1})$$

$$\text{Prob}(S_i) \geqslant n/(n+1); \text{Prob}(\sim S_i) \leqslant 1 - n/(n+1) = 1/(n+1)$$

$$\text{Prob} \sim [S_1 \mathbin{\&} S_2 \mathbin{\&} \cdots \mathbin{\&} S_{n-1}] \leqslant (n-1)\ 1/(n+1) = (n-1)/(n+1) < n/(n+1).$$

Therefore $\sim [S_1 \mathbin{\&} S_2 \mathbin{\&} \cdots \mathbin{\&} S_{n-1}]$ is not probabilistically acceptable.

Now let us observe that Levi's general sets of rationally accepted statements, although they do in fact satisfy the weak consistency principle, do not satisfy the Pairwise Consistency Principle. Consider a three ticket biased lottery, in which ticket 1 has the probability 0.4 of winning, and the tickets 2 and 3 have the probability 0.3 of winning. Relative to the ultimate partition: {Ticket 1 wins, ticket 2 wins, ticket 3 wins}, we will be able to accept the statement 'ticket 1 wins' together with all its deductive consequences. Relative to the ultimate partition {ticket 1 wins, ticket 1 does not win}, we will be able to accept the statement 'ticket one does not win' together with all of its deductive consequences. Thus in the union of the statements accepted relative to each of these ultimate partitions, we will find both a certain statement and its denial (We assume $q = 1$.)

It is interesting to observe that while Levi takes me to task for abandoning the rule of conjunction in my system (and with it the strong deduction principle and the strong consistency principle), the set of rationally accepted beliefs that he comes up with not only abandons all three of these principles (conjunction, strong deduction, strong consistency) but also fails to satisfy the pairwise consistency principle, which my own system, like that of any other purely probabilistic system, does satisfy. It is also interesting to note that the requirements of Rule A, construed locally and not globally, are satisfied by *any* probabilistic rule of acceptance (including mine).

V

Keith Lehrer discusses a purely probabilistic rule of inductive inference, and what amounts to the strong consistency condition. He shudders with horror at what he regards as an abandonment of consistency, and attempts to provide a rule of inductive inference which will satisfy the three strongest principles: the conjunction principle, the strong principle of consistency, and the strong principle of deduction. The principle he comes up with contains as parameters P, an appropriate probability function, e, a body of evidence, and L, a formal language. There is no parameter corresponding to degree of caution, or level of practical certainty. There is no relativization to a given set of hypotheses. The rule is:

RDI: $D(k, e)$ [k is directly inducible from e] if, for any h, if it is not the case that $\ulcorner e \,\&\, k \urcorner \vdash h$, then $P(k, e) > P(h, e)$.

IR: $I(h, e)$ [h is inducible from e] if and only if h is a member of a set I_e such that $I_e = I_1 \cup I_2 \cup \cdots \cup I_m$, [where]
$I_1 =$ the set of hypotheses h_j such that $D(h_j, e)$,
and letting C_n be a conjunction of the members of I_n that is logically equivalent to I_n,
$I_2 =$ the set of hypotheses h_j such that $D(h_j, C_1 \,\&\, e)$
$I_3 =$ the set of hypotheses h_j such that $D(h_j, C_2 \,\&\, e)$
...
$I_m =$ the set of hypotheses h_j such that $D(h_j, C_{m-1} \,\&\, e)$.

We may elucidate the 'logical equivalence' of a statement (C_n) and a set of statements (I_n) in the obvious way: C_n is derivable from the set of statements I_n, and every member of I_n is derivable from C_n. It might be wondered if, for every I_n, there is a finite conjunction of members of I_n from which every member may be derived. In a finite language this is the case; in an infinite language C_n may turn out to be an infinite conjunction. It is possible to prove that the set of statements induced by IR from consistent evidence satisfies the conjunction principle (if h and k are inducible from e by IR, so is their conjunction); that this set of statements is deductively closed (if $h_1, \ldots, h_n$ are inducible from e by IR, and $\ulcorner h_1, \ldots, h_n \urcorner \vdash k$, then k is inducible from e); and that it is strongly consistent (it contains no set of statements $h_1, \ldots, h_n$ such that $\ulcorner h_1, \ldots, h_n \urcorner \vdash \ulcorner k \& \sim k) \urcorner$.

But we have achieved our goal of an inductive rule that satisfies these strong principles at essentially the same cost as that paid for Hintikka and Hilpinen's rule, namely: there is essentially only *one* hypothesis that we may induce from given evidence. Anything else we are allowed to induce will turn out to be merely an implicate of the evidence and that one strongest hypothesis. Indeed the situation for Lehrer's system is even stranger; at a given level we cannot induce all the deductive consequences of the strongest hypothesis we can induce, but only a string of implicates, of which each implies *all* the statements lower in the string.

LEMMA: If h belongs to I_m, then h belongs to I_{m+1}.

PROOF: I_{m+1} is the set of hypotheses h_j such that $D(h_j, C_m \& e)$, i.e., such that for any h, if it is not the case that $C_m \& e \& h_j$ entails h, then $P(h_j, C_m \& e) > P(h, C_m \& e)$. But $C_m \& e$ does entail h, so h is directly inducible from $C_m \& e$.

THEOREM 9: If h and k are directly inducible from e, then either $k \& e$ entails h or $h \& e$ entails k.

PROOF: Suppose that $k \& e$ does not entail h. Then $P(k, e) > P(h, e)$. Suppose that $h \& e$ does not entail k. Then $P(h, e) > P(k, e)$. Therefore either $k \& e$ entails h or $h \& e$ entails k.

Observe that this means that if h and k are any two members of C_n, then either $C_{n-1} \& e \& k \vdash h$ or $C_{n-1} \& e \& h \vdash k$.

(Now, incidentally, we see why the conjunction principle holds. If h and k are directly inducible from e, then so is their conjunction, simply because their conjunction is equivalent, given e, to one of them alone: i.e., either e entails $k \equiv h \& k$, or e entails $h \equiv h \& k$.)

We now come to the main theorem regarding Lehrer's system, which is that there is essentially only *one* statement inducible from e. In order to show this rigorously for infinite languages we must allow ourselves infinite conjunctions; two observations on this move are in order. It seems to be only in pathological cases that the move to infinite conjunctions is required. This move must be allowable in Lehrer's system, since he says to let "C_n be a conjunction of members of I_n that is logically equivalent to

I_n", and in the pathological cases no finite conjunction will do. I suspect, however, that worry about the infinite case is academic.

THEOREM 10:

(a) For every n, there is a statement, k_n, which is either inducible from C_{n-1} and e, or which is an infinite conjunction, each member of which is inducible from C_{n-1}, such that every member of I_n is entailed by k_n and e.

(b) There is a statement, k_s, such that every statement inducible from e is entailed by e and k_s, and such that k_s is inducible from e, or is an infinite conjunction, each member of which is inducible from e.

PROOF: For every two statements in I_n, either the first, in conjunction with e and C_{n-1}, entails the second, or the second, in conjunction with e and C_{n-1}, entails the first. This relation gives a partial ordering of the hypotheses inducible from C_{n-1} and e. Either (as would in general be true) there is a hypothesis k such that every element of I_n is derivable from $k \& e \& C_{n-1}$, and then k is the k_n whose existence is asserted in (a), or else there will be an infinite sequence of hypotheses inducible from $C_{n-1} \& e$, $k_1 \ldots, k_m \ldots$, such that $k_i \& C_{n-1} \& e$ entail k_{i-1}, but $k_{i-1} \& C_{n-1} \& e$ do not entail k_i. In this case we let k_n be the infinite conjunction of the hypotheses k_i. (Observe that in this case, C_n will have to be an infinite conjunction, too. We may, of course take k_n as C_n itself in this case.) Therefore $k_n \& e \& C_{n-1}$ does entail h.

In a similar way, either there is an n^* such that for n and m greater than n^*, C_n and e entail C_m and e and conversely, or else, for every n, C_{n+1} and e entail C_n, but C_n and e do not entail C_{n+1}. In the former case k_n is the k_s of (b). In the latter case, there is an infinite conjunction, C^*, such that every member of C^* is inducible from e, and every finite conjunction of members of C^* is inducible from e, which is such that for any statement h whatever, if h is inducible from e, h is deducible from C^*; namely the conjunction of all k_n.

It is perhaps worth observing that the conjunction principle does not hold for infinite conjunctions. Consider a sequence of hypotheses, $k_1, \ldots, k_n$ each of which, with e, entails the preceding hypotheses, but is not entailed by them (with e). The probability of the conjunction $k_1 \& k_2 \& \cdots \& k_n$, given e, is just the probability of k_n given e. The probability of the infinite conjunction $K_1 = k_1 \& k_2 \& \ldots, k_n \ldots$ is thus just the $\lim_{n\to\infty} P(k_n, e)$. Each of the k_i will be inducible provided that for any h not entailed by $k_i \& e$, $P(k_i, e) > P(h, e)$. But $P(h, e)$ could be $\lim_{n\to\infty} P(k_n, e)$ (since

$P(k_i, e) > \lim_{n\to\infty} P(k_n, e)$), and h might not be entailed even by e and the infinite conjunction of the k_i. Thus the infinite conjunction would not be inducible, even though each of its members was. Thus k_n and k_s may not themselves be acceptable.

VI

The system of Hintikka and Hilpinen and the system of Lehrer are unsatisfactory for essentially the same reason. They boil down to the claim that given a language and a body of evidence, there is essentially just one strongest statement that can be accepted. This approach to induction is global with a vengeance. It suggests that as scientists or even as people we do not induce hypothesis by hypothesis, but that induction consists in principle of inducing at each stage of inquiry – i.e., for each body of evidence e – a single monumentally complex conjunctive statement. Observe that we cannot even consider parts of the complex hypothesis in isolation from other parts. Although the evidence may have the form $e_i \& e_j$, and e_i may be utterly irrelevant to h_j, the fact that h_j is inducible from e_j will have no bearing at all on whether h_j is inducible from $e_i \& e_j$. A hypothesis h_i, not entailed by h_j and e_i and e_j may always turn out to be more probable on $e_i \& e_j$ than h_j is. Indeed, one may wonder if the exceedingly high confirmation of the hypothesis that the speed of light is finite will not preclude the acceptance of *any* hypothesis concerning the cause of cancer, the existence of life on Mars, or the amount of inflation to be anticipated in the coming year.

It may be argued that any global system in which the conjunction principle is satisfied will suffer from these shortcomings. In any such system there will be a statement from which every acceptable statement follows, and which is either acceptable itself, or is an infinite conjunction each finite conjunct of which is acceptable.

THEOREM 11: If the conjunction principle is satisfied for an inductive acceptance rule in the language L, then there is a statement C^* in L, or an infinite conjunction C^* every component of which is in L, such that if h is a finite statement, h is inductively acceptable if and only if h is entailed by C^* and the basic evidence e.

PROOF: Entailment given e provides a partial ordering of the statements in L. Suppose first that there are two statements in L, h_1 and h_2, each of which is inductively acceptable, and neither of which is ranked above the other in the partial ordering. By the conjunction principle their conjunction must also be acceptable, and by the partial ordering, their conjunction must be ranked higher than either one alone. Either there is an acceptable hypothesis of maximum rank (from which all the other hypotheses then follow, given e) or else for every hypothesis h, there is a stronger one h' which, conjoined with e, entails h, but which is not entailed by h conjoined with e. If this is the case, an argument like that of a preceding theorem will give a C^* satisfying the conditions of the theorem.

Given the conjunction principle, as I showed earlier, it also follows that the strong and weak consistency conditions are equivalent, and that the strong and weak deduction principles are equivalent. Thus I think it is appropriate to focus on the conjunction principle as a source of the peculiarities to which such systems as those we have looked at give rise.

It is difficult to give an argument against the conjunction principle, partly because it is so obvious to me that it is false, and partly because it is so obvious to certain other people that it is true. The most persuasive arguments perhaps are those which stem from the last theorem presented; it seems preposterous to suppose that all of our inductive knowledge has to be embodiable in a single fat statement. It seems too limiting to say that I have to believe the conjunction of everything I have a right to believe (there cannot be very much, then, that I have a right to believe), and it seems even more unreasonable to claim to have a right to believe the conjunction of everything I have a right to believe. Although I claim to have good reasons for believing every statement I believe, I claim also to have good reasons for believing that some of those statements are false. I think both of those claims are perfectly sound; and if they are, the conjunction principle is false.

VII

The system of rational beliefs I have developed elsewhere accepts the failure of the principle of conjunction. It is an attempt at a global theory, and for a global theory the conjunction principle seems flatly false. Having abandoned the principle of conjunction, it becomes possible to distinguish strong and weak deductive closure, strong, weak, pairwise and n-wise consistency. In what follows I will briefly characterize a simplified version of the original system, freed, of course, from the original inconsistency.

We begin (as always) with a language L; we suppose it to contain a set theory, terms denoting operations, properties, relations, etc., which may be of a theoretical as well as of an observational character. We let B denote a set of statements that are accepted as evidence or background information. In a given context B represents the basic rational corpus, or body of knowledge. We suppose, by hypothesis, that B is pairwise consistent, and satisfies weak deductive closure. B thus contains all the theorems of our language. To be sure, if we pick the wrong axioms for set theory – inconsistent ones – in our language, the set B will be empty. But we must always suppose, so long as we know no better, that the language we speak is consistent. We do not suppose that we believe in any active or behavioristic sense every statement in B; rather we say that the contents of B are what we are *committed* to believing.

AXIOM I: $S \in B \supset \sim nS \in B$ ('nS' denotes the denial, in L, of S).

AXIOM II: Thm S cd T & $S \in B \supset T \in B$ ('S cd T' denotes the conditional whose antededent is S, and whose consequent is T; 'Thm S cd T' says that the statement S cd T is a theorem of L).

There is no need here for defining the probability relation; there are certain properties of that relation I shall refer to, which I shall state as axioms. It should be observed that on the basis of a definition of probability like that I have provided elsewhere, these axioms turn out to be theorems. Probability I take to be relative to a body of knowledge or rational corpus B; it is a relation that holds between a statement S, the rational corpus B, and a pair of fractions p and q. We say, relative to B, the probability of S is the pair of fractions (p, q), and we symbolize this assertion: 'Prob_B (S, p, q)'. It should be observed, not as part of the formal development here, but simply as background information, that on the definition I have offered, every probability statement is based on a *known* statement concerning relative frequencies, i.e., that if the relation $\text{Prob}_B(S, p, q)$ holds, there is as a member of B some corresponding statistical statement asserting that a certain relative frequency or measure lies between the ratios denoted by p and q. The properties of probability that we shall need are the following:

AXIOM III: $\text{Prob}_B(S, `p', `q') \supset \text{Prob}_B(nS, `1-q', `1-p')$.

AXIOM IV: [Thm S cd T & $\text{Prob}_B(S, p, q)$ & $\text{Prob}_B(T, p', q')$] $\supset$ Thm p' gr p ('p' gr p' denotes the statement in abbreviated notation consisting of the fraction p', followed by '$>$', followed by the fraction p).

AXIOM V: $S \in B \supset \mathrm{Prob}_B(S, \text{'1/1'}, \text{'1/1'})$.

Our final axiom requires an auxiliary notion, that of a biconditional chain in B. Since we have not assumed deductive closure in B, it is perfectly possible that $S\,b\,T$ (the biconditional whose antecedent is S and whose consequent is T) is a member of B and that $T\,b\,R$ is a member of B, when $S\,b\,R$ is not a member of B. But we want to say that S and R are related by a biconditional chain in B anyway. In general we shall say that S and R are related by a biconditional chain in B, in symbols, $S\,bc_B\,T$, when $S\,b\,T$ belongs to every set of statements containing $P\,b\,Q$ whenever $P\,b\,Q$ belongs to B, and containing $P\,b\,R$ whenever it contains both $P\,b\,Q$ and $Q\,b\,R$. Formally:

DEFINITION 1: $S\,bc_B T = \mathrm{df}\,(K)((P)\ (Q)\ (R)\ ((P\,b\,Q \in B \supset P\,b\,Q \in K)\,\&\,(P\,b\,R \in K \,\&\, P\,b\,Q \in\ \supset P\,b\,Q \in K)) \supset S\,b\,T \in K)$.

The final axiom simply says that any two statements related by a biconditional chain have essentially the same probability.

AXIOM VI: $(S\,bc_B\,T \,\&\, \mathrm{Prob}_B(S, p, q) \,\&\, \mathrm{Prob}_B(T, p', q')) \supset (\mathrm{Thm}\ p\ \mathrm{id}\ p' \,\&\, \mathrm{Thm}\ q\ \mathrm{id}\ q')$ ('x id y' denotes the statement [in abbreviated notation] consisting of x followed by '=' followed by y).

We are now in a position to characterize the set of statements B_r which may be induced from B, at the level r. Note that we cannot simply include a statement in B itself on the grounds that its probability is at least r, unless we take r to be 1, for the probability of S, relative to B, can be less than one only if S is not a member of B.

THEOREM 12: $(\mathrm{Prob}_B(S, \text{'}p\text{'}, \text{'}q\text{'}) \,\&\, \mathrm{Thm}\ \text{'}1 > p\text{'}) \supset \sim S \in B$ (Axioms V and VI).

Let us define B_r to be the set of statements whose probability, relative to B, is at least r, where r denotes a ratio greater than a half.

DEFINITION 2: $B_r = \mathrm{df}\,\{S\colon \mathrm{Prob}_B(S, p, q) \,\&\, \mathrm{Thm}\ p\ \mathrm{gr}\ r\}$.

B_r is thus a set of statements accepted on purely probabilistic grounds. We can show that B_r satisfies the weak consistency principle, the weak deduction principle, and the pairwise consistency principle, which, recall, failed for Levi's general system.

THEOREM 13: $(S \in B_r \,\&\, \mathrm{Thm}\ S\ \mathrm{cd}\ T) \supset T \in B_r$ (Axiom IV, D-3).

THEOREM 14: $\sim (\exists S)\ (S \in B_r \,\&\, (T)\ (\mathrm{Thm}\ S\ \mathrm{cd}\ \mathrm{T}))$.

PROOF: If (T) (Thm S cd T), then Thm nS. Thus $nS \in B$. $\mathrm{Prob}_B(nS, \text{'1'}, \text{'1'})$. By axiom III, then, $\mathrm{Prob}_B(S, \text{'0'}, \text{'0'})$. By the (hypothetical) consistency of B, we have $\sim$Thm '0' gr r, and thus $\sim S \in B_r$.

THEOREM 15: (Pairwise Consistency and n-wise consistency for n such that $n/(n+1) < r$)

$$(S)\,(S \in B_r \supset \sim nS \in B_r) \text{ (Axiom III).}$$
$$(S_1)\,(S_2)\ldots(S_{n-1})\,[(S_1 \in B_r \,\&\, S_2 \in B_r \,\&\cdots\&\, S_{n-1} \in B_r) \supset \sim n(S_1 \text{ cj } S_2 \ldots \text{ cj } S_{n-1}) \in B_r]$$

(where S cj T represents the conjunction of S and T).

Lehrer makes a point of the consistency of what is induced with the evidence on the basis of which it is induced. This condition is satisfied here for single statements given that B is deductively closed, which may not be an unreasonable supposition for 'observation statements'.

THEOREM 16: $(S)(B \vdash S \supset S \in B) \supset (S)\,(S \in B_r \supset \sim (B, S \vdash \text{contradiction}))$.

PROOF: If $B, S \vdash$ contradiction, then $B \vdash nS$. By the hypothesis of the theorem, then, $nS \in B$, and $\text{Prob}_B(S, \text{'0'}, \text{'0'})$, and $\sim S \in B_r$.

As I pointed out earlier, there are parts of any system like this which are strongly consistent and deductively closed. The relation between statements S and T, Thm S cd T, provides a partial ordering of the elements of B_r. Let us define a strongest acceptable statement, in symbols STR, to be a statement such that no other statement in B_r bears the relation in question to it, unless it also bears that relation to the statement. Thus:

DEFINITION 3: $STR_{B_r}\, S = df\, S \in B_r \,\&\, (T)\,((T \in B_r \,\&\, \text{Thm } T \text{ cd } S) \supset \text{Thm } S \text{ cd } T)$.

The set of consequences of any strongest statement in B_r satisfies the strong deduction principle, the strong consistency principle, and (!) even the conjunction principle.

THEOREM 17: $STR_{B_r}\, S \supset [(T)\,(\text{Thm } S \text{ cd } T \equiv T \in A) \supset (\sim A \vdash \text{contradiction} \,\&\, (R)\,(A \vdash R \supset R \in A) \,\&\, (R)\,(T)\,((R \in A \,\&\, T \in A) \supset R\; cj\; T \in A))]$.

Perhaps, in these terms, it is the fact that most of our beliefs are not strongest beliefs that has led people to feel that our beliefs belong to systems of beliefs which satisfy the conjunction principle.

There are a number of other theorems we can prove. For example, we can prove that if a complex statement S is like a complex statement T, except for containing occurrences of the statement P where T contains occurrences of the statement Q, and if P and Q are connected by a biconditional chain in B, then S will be a member of B_r if and only if T is a member of B_r. In a similar fashion it is possible to define the concept of an identity chain in B. Then it is possible to prove that if S is like T except for containing occurrences of the term p where T contains occurrences of the term q, and p and q are connected by an identity chain in B, then S will be a member of B_r if and only if T is a member of B_r. Even without complete deductive closure, there is a lot that can be shown to hold in B_r.

The issue is only whether or not there is a single strongest statement in B_r – i.e., whether there is a statement S^* such that (T) $(T \in B_r \supset \text{Thm } S^*$ cd $T)$. When stated thus baldly, the answer is obvious; it is gratuitious to suppose that there is any such statement. Indeed the supposition that there is is one of the secondary symptoms of the disease I have called conjunctivitis.

University of Rochester

BIBLIOGRAPHY

[1] R. B. Braithwaite, *Scientific Explanation*, Cambridge 1953.

[2] Gilbert Harman, 'Detachment, Probability and Maximum Likelihood', *Nous* **1** (1967) 401–11.

[3] Carl Hempel, 'Deductive-Nomological vs. Statistical Explanation', in *Minnesota Studies in the Philosophy of Science*, III (ed. by H. Feigl and G. Maxwell), Minneapolis 1962.

[4] Jaakko Hintikka, 'Towards a Theory of Inductive Generalization', in *Proc. 1964 International Congress for Logic, Methodology, and Philosophy of Science* (ed. by J. Bar-Hillel), Amsterdam 1965.

[5] Jaakko Hintikka, 'Distributive Normal Forms in First-Order Logic', in *Formal Systems and Recursive Functions* (ed. by Crossley and Dummett), Amsterdam 1965.

[6] Jaakko Hintikka, 'A Two-Dimensional Continuum of Inductive Methods', in *Aspects of Inductive Logic* (ed. by J. Hintikka and P. Suppes), Amsterdam 1966.

[7] Jaakko Hintikka and Risto Hilpinen, 'Knowledge, Acceptance, and Inductive Logic', in *Aspects of Inductive Logic* (ed. by J. Hintikka and P. Suppes), Amsterdam 1966.

[8] Richard Jeffrey, *The Logic of Decision*, New York 1965.

[9] Henry E. Kyburg, Jr., *Probability and the Logic of Rational Belief*, Middletown 1961.

[10] Henry E. Kyburg Jr., 'Probability and Randomness', *Theoria*, **29** (1963).

[11] Henry E. Kyburg Jr., 'Probability, Rationality, and Rule of Detachment', in *Proc. 1964 Congress for Logic, Methodology and Philosophy of Science* (ed. by J. Bar-Hillel), Amsterdam 1965.

[12] Keith Lerher, 'Induction, Reason, and Consistency', forthcoming in *Minnesota Studies in the Philosophy of Science*.

[13] Isaac Levi, *Gambling With Truth*, New York 1967.

[14] Wesley Salmon, private correspondence.

[15] Raimo Tuomela, 'Inductive Generalization in an Ordered Universe', in *Aspects of Inductive Logic* (ed. by J. Hintikka and P. Suppes), Amsterdam 1966.

REFERENCES

* Much of the research on which this paper is based has been supported by the National Science Foundation, through grants 708, 1179, and 1962.

1 Except in one of Keith Lehrer's systems, described in this volume.

2 A muddier version of this argument was presented in [11]; a cleaned-up version is mentioned by Harman in [2].

3 These principles are not essential parts of the Bayesian or Classical statistical theory. One can develop the theory of statistical inference without considering the question of acceptance one way or the other. The classical theory requires us to *reject* certain hypotheses, but it is hardly necessary to point out (as statisticians of this persuasion inevitably do) that to reject a statement is not (necessarily) to accept it. Bayesian theory is sometimes coupled with a philosophy according to which one *never* accepts any *hypothesis*.

4 The test interval for '$p \in (a, f)$' will include the closed interval $[a, f]$ and the test interval for '$p \in (f, b)$' will include the closed interval $[f, b]$.

GILBERT H. HARMAN

INDUCTION*

A Discussion of the Relevance of the Theory of Knowledge to the Theory of Induction (with a Digression to the Effect that neither Deductive Logic nor the Probability Calculus has Anything to Do with Inference)

I

In 1963 Edmund Gettier demonstrated that knowledge is not simply justified true belief[1]; and what has been learned in the resulting discussion[2] has important implications for a theory of reasoning. This paper describes some of those implications; more generally, it attempts to show how theories of knowledge and reasoning must be adapted to each other if one is to achieve a unified theory of both.

An obvious connection between one's theory of knowledge and one's theory of reasoning is that one can take reasoning to be warranted or valid if it could give a person knowledge. For example, a detective comes to know who the murderer is by reasoning from circumstantial evidence: in such a case his reasoning can be said to be valid or warranted. Whether a person knows something by reasoning depends (in part) on whether his reasoning justifies his belief. In the language of inductive logic, knowledge depends on whether reasoning justifies acceptance of one's conclusion. Epistemologists refer to principles that warrant belief where logicians refer to rules of acceptance.

If we thus approach inductive logic from the theory of knowledge, we will want it to provide a *strong rule of acceptance*. Roughly speaking, such a rule tells one that, given certain evidence, one may infer, accept, or believe nonprobabilistic conclusions which may be used as part of the evidence in further reasoning or inference. That is oversimplified, since sometimes one should *reject* something previously accepted. More precisely, prior to inference one accepts a set of propositions which serve as premises or evidence; inference leads to a modification of the set either by the acceptance of further propositions or by the rejection of propositions previously accepted; and one can then use the revised set as a basis for future inference.

A person can come to know something by inference only if rules of acceptance authorize him to accept it. The following test of proposed rules of acceptance therefore suggests itself: Can a person come to know something by reasoning that proceeds according to the rules in question? A negative answer counts against the rules; a positive answer counts in their favor. The test is not conclusive, since rules of inference are not the only conditions of inferential knowledge; but it proves nevertheless to be very useful in the study of inductive reasoning.

We can apply the test only if we can determine when a person knows something. Here we must rely on ordinary judgments not biased by philosophical theory. Naturally, disagreements sometimes arise; but we must not exaggerate their importance. The extensive recent literature on the definition of 'know' contains wide disagreement about principle but remarkable agreement on cases; and it is much easier to reach agreement about instances of knowledge than about principles of inference or particular inferences.

For example, defenders of subjective probability argue against rules of acceptance, while other philosophers argue in favor of such rules. Although one cannot resolve that issue by direct appeal to intuitions about inference, one can give an indirect argument for rules of acceptance that appeals to intuitions about knowledge:

> We are ordinarily willing to say that a smart detective can come to know who the murderer is by reasoning from circumstantial evidence. If what we say is true, such a detective comes to know something by inductive inference. If knowledge involves warranted belief or acceptance, he is warranted in believing or accepting his conclusion. So, if inductive logic is what emerges from a study of the ordinary use of 'know' and 'believe' it must formulate one or more rules of acceptance.

Although this argument is not conclusive, it provides a case to be met by anyone who denies that there are rules of acceptance. For my part, I see only two ways in which one could hope to meet that case, neither very promising. First, certain subjective probability theorists might try to account for ordinary usage on the assumption that a person can know, without being warranted in accepting what he knows, as long as he is warranted in believing it to a certain degree. Thus they might argue that a person never simply believes anything but has instead some degree of belief in it; and they might claim that inductive logic consists in the application of the probability calculus to ensure that one's degrees of belief are 'coherent'. However, given the complex nature of intuitions about knowledge to be described below, it is difficult to see how such an approach could successfully account for ordinary usage.

Alternatively, skeptics might argue that this very complexity is a sign of conceptual confusion.[3] They might claim that it shows that ordinary usage is incoherent. But such a conclusion would be premature. We can only discover whether sense can be made of ordinary notions if we try to make sense of them. At the present time, we must take ordinary usage seriously. So we must take seriously the idea that there are inductive rules of acceptance.

II

It is well known that strong rules of acceptance cannot be purely probabilistic. They cannot simply tell one to accept any proposition that is more probable than some fixed amount, since that would permit one to infer, for some lottery, both that any ticket will lose and that some ticket will win. It is true that if probabilities are conditional on the total evidence, no contradiction would result, since whenever one inferred a particular ticket was going to lose that would affect the probability that any other ticket was going to lose. (Recall that conclusions gotten by *strong* rules must be added to the evidence.) But, even if they would not lead one into contradiction, purely probabilistic rules would have at least two serious defects: (i) the conclusions that could be drawn in accordance with such rules would depend on the order in which they were drawn; and (ii) for a sufficiently large lottery such rules would permit one to accept a set of conclusions the antecedent probability of whose joint truth was extremely small.

The lottery paradox indicates that rules of acceptance are not purely probabilistic. The same conclusion follows from examination of the ordinary use of 'know'. Gettier points out that one can validly infer a true conclusion from premises one knows to be correct yet fail to know that one's conclusion is true. We can account for Gettier's point only if we suppose that rules of acceptance are not purely probabilistic.

For example, Jones knows that he has seen his officemate Smith driving a Ford; he knows that he has heard Smith say that he owns a Ford; he knows that he has seen an automobile registration slip made out in

Smith's name; etc. From such premises, all of which he knows to be true, he infers 'Someone in the office owns a Ford'. We may suppose that he is warranted in making that inference; and, if we suppose that Smith did own and continues to own a Ford, we may suppose that Jones knows that someone in the office owns a Ford. On the other hand, we can also suppose that although Jones is still warranted in making his inference, in fact Smith does not own a Ford and therefore Jones does not thereby come to know that someone in the office owns a Ford. In other words, we can imagine two cases in which Jones is equally warranted in drawing a true conclusion from the same premises he knows to be correct, such that he comes to know in the first case but not in the second.[4]

It is natural to try to explain what goes wrong in the second case by referring to the way in which Jones must reach his conclusion.

> Logically speaking, Jones can infer that someone in the office owns a Ford only because he can first infer that his officemate Smith owns a Ford. He does not know that his final conclusion is true unless that intermediate conclusion is also true.

If such an explanation is even roughly correct (and I think it is), inductive inference cannot use purely probabilistic rules of acceptance. If he could accept any conclusion whose probability exceeded a certain amount, Jones would not have to infer his intermediate conclusion before reaching his final conclusion. The probability of that final conclusion is at least as great as the probability of his intermediate conclusion on the original evidence. Any purely probabilistic rule that would license direct acceptance of the intermediate conclusion would also license direct acceptance of the final conclusion. So, if rules of acceptance were purely probabilistic Jones' inference would not require any intermediate step, and the natural explanation of his lack of knowledge in the second case could not be correct.

One can appreciate this argument for nonprobabilistic rules of acceptance only after one has attempted to give a coherent account of ordinary usage. One quickly discovers that any plausible theory involves complicated proposals about reasoning.

For example, consider Alvin Goldman's 'causal theory of knowing'.[5] He suggests one knows only if there is the proper sort of 'causal connection' between one's belief and what one knows. For example, one perceives that there has been an automobile accident only if the accident is relevantly causally responsible, via one's sense organs, for one's belief that there has been an accident. Similarly, one remembers doing something only if having done it is relevantly causally responsible for one's current memory that one did it. Sometimes the fact one knows begins a causal chain that leads to his belief. At other times, the connection is more complicated. Thus, when Jones knows that someone in the office owns a Ford, Smith's past ownership is causally responsible for the evidence Jones has and also responsible (at least in part) for someone's present ownership. Here the relevant causal connection consists in there being a common cause of the belief and the state of affairs believed to exist.

General knowledge does not fit into such a simple framework. That all ravens are black neither causes nor is caused by the existence of the particular black ravens examined when one comes to know that all ravens are black. Goldman handles such examples by counting certain logical connections as relevant 'causal connections'. One's belief that all ravens are black is, in an extended sense, relevantly causally connected to the fact that all ravens are black. One's evidence causes one's belief and is logically entailed by what what one believes.[6]

It is obvious that not every causal connection in this extended sense is relevant. Any two states of affairs are logically connected simply because both are entailed by their conjunction or because both entail their disjunction. If every such connection were relevant, the analysis would have us identify knowledge with true belief, since there would always be a relevant causal connection between any state of true belief and the state of affairs believed in.

Goldman says that when knowledge is based on inference, relevant causal connections must be 'reconstructed' in one's inference. Jones knows that someone in the office owns a Ford only if his inference reconstructs the relevant causal connection between evidence and conclusion. We may learn something about relevant causal connections if we discover what connections are inferentially reconstructable. A theory of knowledge requires a theory of inference.

Recall the two cases. In the first Smith owns the Ford; in the second someone else in the office does. Jones knows only in the first case that someone in the office owns a Ford. I mentioned above that the natural explanation is that Jones must first infer the intermediate conclusion,

'Smith owns a Ford', and does not know his final conclusion is true unless that intermediate conclusion is true. If we are to join this natural explanation with Goldman's suggestion, that Jones knows only if his inference 'reconstructs' a relevant causal connection, we must turn Goldman's theory into a theory of inference.

As a first approximation, let us take Goldman's remarks about 'causal connections' literally, forgetting for the moment that they include logical connections. Then let us transmute his causal theory of knowing into the theory that inductive conclusions always take the form *X causes Y*, where further conclusions are reached by additional steps of inductive or deductive reasoning (in particular one may deduce either *X* or *Y* from *X causes Y*). This causal theory of inference provides the following explanation of why a person who knows that his conclusion is true must be right about an appropriate causal connection.

> A person knows by inference only if his intermediate conclusions (or 'lemmas') are true. That is, his inference must satisfy 'the condition that the lemmas be true'. Since he can legitimately infer his conclusion only if he can first infer certain causal statements, he can know only if he is right about the causal connection expressed by those statements.

Jones infers that his evidence is a causal result of Smith's past ownership of the Ford. From that he deduces that Smith has owned a Ford. Then he infers that this past ownership has been causally responsible for present ownership; and he deduces that Smith owns a Ford. Finally he deduces that someone in the office owns a Ford.

We must add that a person's knowledge can be 'based on' particular reasoning even if his conscious reasoning leaves out steps, contains extra material, etc., sometimes even if he is not conscious of reasoning at all. Inferential knowledge requires only that *there be* reasoning from known premises to a man's conclusion such that (a) the reasoning is warranted, (b) it satisfies the condition that the lemmas be true – i.e. all intermediate (and final) conclusions are true – and (c) his belief in his conclusion is in part 'based on' that reasoning.[7]

A better account of knowledge and inference replaces *causal connection* with *explanatory connection*. Induction is conceived as inference to the best (i.e. most plausible) of competing explanatory statements. Inference to the best of competing causal explanations is only a special case.

This avoids Goldman's *ad hoc* account of knowledge of generalizations. Although there is no causal relationship between a generalization and those observed instances which provide us with evidence for the generalization, there is an obvious explanatory relationship; and (other things equal) one can infer a generalization only if it provides the most plausible way to explain one's evidence. Indeed, one often infers generalizations that explain but do not logically entail their instances, since they are of the form, *In circumstances C, X's tend to be Y's.* Such generalizations may be inferred if they provide a sufficiently more plausible account of observed instances than alternatives. Instances may be inferred from them if the explanatory claim *that X's tend to be Y's will explain why the next X will be Y* is sufficiently more plausible than competitors such as *interfering factor Q will prevent the next X from being a Y.*[8]

Statistical inference, and knowledge obtained from it, is also better explicated via the notion of statistical explanation than via the notion of cause or logical entailment. A person may infer that a particular coin is biased because that provides the best statistical explanation of the observed fraction of heads. His conclusion explains his evidence but neither causes nor entails it.[9]

So we are led to construe induction as inference to the best explanation, or more precisely as inference to the best of competing explanatory statements. The conclusion of any single step of such inference is always of the form *X explains Y*, from which one may deduce either *X* or *Y*. Inductive reasoning is thus seen to consist in a sequence of explanatory conclusions; its principles are principles for deciding among competing explanatory statements.

We have been led to this conception of induction in an attempt to account for examples that Gettier offers against the analysis of knowledge as justified true belief. We have tried to find principles of inference which, together with the condition that the lemmas be true, would explain Gettier's deviant cases. Purely probabilistic rules were immediately seen to be inadequate. Goldman's causal theory of knowing, which promised answers to some of Gettier's difficulties, suggested a causal theory of induction: inductive inference as inference to the best of competing causal statements. Our present version is simply a modification of that, with *explanatory* replacing *causal*. Its strength lies in the way it enables one to account for a variety of examples of inferential knowledge by way of the condition that the lemmas be true.

Unfortunately, it also oversimplifies induction. The time has come to say how it has done that and how the theory of inference can be made more adequate. After that we shall discover how complexity in the theory of inference leads to complexity in the theory of knowledge. The next section focusses on inference; the following (and final) section discusses what modifications are required in one's theory of knowledge when one has a more adequate theory of inference.

III

The following three considerations indicate that inductive reasoning cannot be analysed simply as a sequence of conclusions of the form *X explains Y*. First, none of these conclusions can be evaluated apart from the others, since an important test of good inductive inference is the plausibility of the final overall result. Second, induction does not always lead to new conclusions, since it occasionally leads to the rejection of something previously accepted. Third, inference to the best explanatory statement would require deductive inference[10]; but there is no such thing.

In order to clarify these points, let us suppose a person wants to understand as much as possible about the world and also accepts a principle of laziness or conservatism. As a result he tries to do as little as he can in order to bring his view as close as possible to an explanatorily complete and coherent whole. He accepts a hypothesis if, compared with competitors, it fits relatively easily into his total explanatory picture. He rejects a hypothesis if it would produce too much explanatory complication or too many loose ends in his total view. What such a person does constitutes inductive reasoning.

Induction may be conceived in the large or in the small. Its conclusion may be taken to be either one's total resulting view or whatever has been added to one's initial view. To say that induction is not simply induction to the best of competing explanatory statements is to say that it must be conceived in the large as well as in the small. Thus, consider the points alluded to a moment ago.

(1) Consider a sequence of inductive conclusions of the form *X explains Y*: 'None of these conclusions can be evaluated apart from the others,

since an important test of good inductive inference is the plausibility of the final overall result.' For example, consider a version of the lottery paradox in which one attempts to infer *N* different explanations, each (when considered in isolation) the best explanation of what it explains, even though one knows that at most $N-1$ of the explanations are correct. In such a situation one must attempt to discover the best single overall account, where competing accounts contain various ones of the *N* smaller explanations. One should not attempt to infer the smaller explanations considered apart from each other. Versions of the lottery paradox appear to arise only when inductive reasoning, viewed in the small, is conceived as a series of isolated inferences. When we view it in the large, we see how paradox is avoided; for the small inductions are warranted only to the extent that a single large induction (containing their conclusions) is warranted; and the lottery paradox cannot arise if in the end one makes only a single inference. We can avoid the lottery paradox, then, but only if we give up the idea that induction is inference to the best of competing local explanatory statements.

(2) 'Induction does not always lead to new conclusions, since it occasionally leads to the rejection of something previously accepted.' For example, an attempt to discover the best possible explanation of the evidence can lead one to reject part of it as spurious. No novel principle is needed: one rejects a view for its tendency to promote explanatory complication. But such an inference does not simply add something to one's view, since it subtracts something from that view. We must conceive this inference in the large, where one's 'conclusion' is one's total set of beliefs, now missing a statement it previously contained. It follows that we cannot view the inference as inference to the best explanatory statement.

(3) 'Inference to the best explanatory statement would require deductive inference; but there is no such thing.' The relevant sort of inference modifies one's set of antecedent beliefs by adding and sometimes subtracting beliefs. So called deductive 'rules of inference', such as *modus ponens*, cannot be construed as principles of warranted inference in this sense. *Modus ponens* does not say that if one believes both *P* and *If P, then Q*, one may infer and believe *Q*; for sometimes one should give up *P* or *If P, then Q* rather than believe *Q*.

This last point is important enough for elaboration. Suppose a person discovers his beliefs are inconsistent. What should he do? Principles of inference should apply as much to this situation as to any other. However, deductive logic gives him no help at all. According to it, everything 'follows' from a set of mutually inconsistent propositions. But we can all agree that, when a person discovers his views are inconsistent, he should not respond by believing everything he can think of (self contradictions included). Instead, he should follow his usual inductive practice, which is to make the least change in his beliefs that leads to the most explanatory coherence.

Deductive logic is not a theory of inference. It is (i) a specification of certain very general truths and (ii) a theory of deductive relations like entailment. It has no more to do with reasoning than mathematics, physics, or any other very general theory does.[11]

Therefore, if we want inductive logic to provide a theory of inference that says something about the justification of belief, it is a serious mistake to say that the difference between deduction and induction is that deduction is truth preserving whereas induction is not. Unfortunately, it is a mistake that philosophers have often made, with disastrous philosophical consequences. For example, this mistake is what is behind the idea that one justifies a belief by deriving it from more basic beliefs, an idea which leads to the skeptical suggestion that one can never be justified in believing anything, since one's most basic beliefs are ultimately unjustifiable. In ethics, similar considerations have been responsible for relativism and noncognitivism.

The same mistake encourages inductive logicians to overemphasize the probability calculus. If one believes that deductive 'inference' has something to do with deductive relations like entailment, one will think that inductive inference must have something to do with inductive or probabilistic relations of the sort specified by probability theory. This can be very misleading.

For example, it is a truism that inductive conclusions must be assessed with respect to one's total evidence. But philosophers who believe in the probability calculus claim this means e.g. that inductive conclusions ought to have a high probability conditional on the total evidence and that one should reject hypotheses which have a low probability conditional on the total evidence. That might be a truism too, except for the technical meaning given the notion *conditional on the total evidence*.[12] We shall soon see that, so interpreted, what they say is wrong; but even if they were right about probability, they would still not have captured the full meaning of the principle of total evidence. For that principle derives from the fact that induction infers a total explanatory account.

Consider how one uses a principle of total evidence in resolving the last version of the lottery paradox discussed above. N explanations were assumed inferable when considered in isolation, although the evidence also insured that only $N-1$ of them could be true. If one could appeal to total evidence only in order to assign probabilities, one would have no way to avoid paradox since the relevant probability of each explanation can be assumed to be $(N-1)/N$ (which for suitable choice of N is as high as one might wish). When one appeals to total evidence in resolving this paradox, one is not primarily concerned with probabilities; rather one is pointing out that induction conceived in the large is inference to the best overall story and that an explanation can be inferred only if it is part of an inferable total account.

Furthermore, the relevant probabilities are not those in the technical sense conditional on the total evidence. No matter how evidence is conceived, induction occasionally tells one to reject some of it, and that is to reject something that has probability 1 on the total evidence.[13]

Some philosophers[14] would attempt to evade this argument as follows. They would distinguish propositions accepted as evidence from other propositions accepted but not as evidence. Conditional probability on the total evidence in this restricted sense would be relevant in deciding questions of simple acceptance or rejection. In deciding whether to change the propositions accepted as evidence, one would appeal to (unspecified) considerations that are different from those relevant to mere acceptance or rejection. In particular, they would not refer to conditional probability on the total evidence.

But why suppose that principles relevant to acceptance or rejection as evidence would not also account for ordinary inference? In both cases one should make the least change that would maximize explanatory coherence. The distinction between what one accepts as evidence and whatever else one accepts is an untenable dualism, a metaphysical article of faith.

Perhaps reasoning is concerned with subjective probability[15], but it is important to see that one fixes such probability by appeal to explanatory plausibility rather than vice versa. For example, suppose one wishes to determine the subjective probability of heads on a single toss of a given coin. A good method[16] is to consider what one would say about possible outcomes of ten tosses of the coin. If one obtained six heads and four tails, would one be more inclined to say this was the result of random tosses with a fair coin or with a coin biased slightly towards heads? If one would more readily accept the explanation involving the fair coin, one should assign that hypothesis a higher initial subjective probability.... And so on, as dictated by judgments of plausibility in conjunction with subjective probability theory.

(Much depends on how the coin might have come to possess a particular statistical probability of heads. If one believes the coin has been randomly selected from those in circulation, one will feel confident that it is fair. One's suspicions are raised if one thinks the coin has been specially selected or prepared. One can evaluate such suspicions by asking oneself why the owner of the coin might have thus spent his time. If one knows the owner carefully prepared the coin to have a statistical probability of heads that falls somewhere between .5 and .7, one tends to imagine he has selected the exact figure at random.... In each case one arrives at an estimate of subjective probability by considering the plausibility of various explanations.)

One has degrees of belief only because one has beliefs. One's degree of belief in a proposition has something to do with how easily one could modify one's total set of beliefs in order to accept or reject that proposition. Perhaps we should speak of a *subjective plausibility function*, which could be derived from judgments of explanatory plausibility. Subjective probability would depend on one's beliefs and one's subjective plausibility function. But that is only a suggestion. Whether or not it is right, a great deal of study is needed here.

Let me pause to say where we are. We began by considering how the theory of knowledge and the theory of inference were related. This led us to seek a theory of inference that, together with the condition that the lemmas be true, would account for Gettier-type counter examples to traditional analyses of knowledge. Goldman's causal theory of knowing suggested a related theory of inference; and, after considering various improvements, we arrived at inference to the best explanatory statement. In the present section, we discovered that our theory was oversimplified and that, ideally, one infers a total explanatory account. We must now consider how the theory of knowledge is affected by this change in the theory of reasoning.

IV

First, notice that the condition that the lemmas be true is appropriate only to induction conceived in the small. Conceived in the large, induction has as its conclusion one's total resulting view; but a person can know something by inference even though not everything he believes is true. Let us call the total resulting view a *large conclusion* and the set of beliefs newly acquired by inference a *small conclusion*. Furthermore let us say that such a conclusion is true if all of its members are true. Then we can state the condition that the lemmas be true as follows:

> A person comes to know *h* by inference only if there is at least one warranted inference (on which his belief is based[17]) to a true small conclusion that contains *h*.

A similar point must be made about premises. It is prima facie plausible to suppose one can know by inference only if one already knows one's premises to be true. On the other hand, the validity of inference depends on one's total initial view, and presumably one initially believes things one does not know to be true. So we must require, not that one's inference be from known premises, but rather that it remain warranted when its premises are restricted to things one knows. Combining that with the condition that the lemmas be true yields the following important principle.

(P) A person comes to know *h* by inference only if there is at least one warranted inference (a) on which one's belief in *h* is based, (b) to a true small conclusion that contains *h*, and (c) which remains warranted when its premises are restricted to things one knows.

Principle (P) accounts for a variety of Gettier-type examples, perhaps for all. I once thought it could not handle cases in which one fails to know something because of the existence of evidence one does not possess[18]; but I now think that I was wrong. Principle (P) can handle such cases if an assumption is made about inference. Let me elaborate.

Smith reads in the paper about the assassination of some famous person. The story was written by a reporter who witnessed the event and immediately cabled his paper which put out an extra. Ordinarily in such circumstances Smith would know the assassination had occurred. But in this case for political reasons the story has been denied by seemingly authoritative sources. Although the denials occurred too late to stop publication of the paper with the true story, they have received wide circulation and have been generally accepted. However, Smith believes the story in the paper, since he has not heard of the denials. In such a case, Smith does not know. I thought that (P) could thus be satisfied in a Gettier-type case. Smith's inference remains warranted when restricted to known premises; and it seemed to me that his small conclusion would be completely true. Therefore, I concluded that (P) cannot account for all such cases.

I should point out that it remains obscure how to formulate an explicit principle expressing the relevance of evidence one does not possess. Smith's inference would not have been warranted had he known about the denials; and that suggests the following principle:

> If one comes to know by inference, the inference from known premises must remain warranted when any true proposition is added to the premises.

But that principle is too strong. The law of averages ensures that for any conclusion there are many true propositions which, considered in isolation, could be thus used as evidence against that conclusion. We must distinguish the *special* evidence that undermines a claim to knowledge from other evidence. One's inference must remain valid when any true proposition stating special[19] further evidence is added to the premises or in short:

(Q) There must exist no special counter evidence to one's conclusion.

The problem is, then, to show how (P) can account for (Q). The key to a solution is given by Sosa[20], who points out that a principle analogous to (Q) holds of warranted inference.

(R) One may infer a particular conclusion only if one is also warranted in inferring that there is no special counter evidence to that conclusion.

Thus Smith is warranted in believing in the assassination only if he is also warranted in believing that there exist no seemingly authoritative denials, etc.

Notice that (R) would give rise to an infinite regress if the second inference mentioned were different from the first. The regress could not arise if the following principle were right:

(S) All (small) inductive conclusions contain the (self-referential) claim that there is no special counter evidence to that whole conclusion.

(R) follows from (S). Furthermore (Q), which expresses the relevance of evidence one does not possess, follows from (R) and (P), the principle that is to account for all Gettier-type cases.

(S) and (R) are principles of the theory of reasoning. If we assume one of them is true, we can use (P) to account for Gettier-type examples involving evidence one does not possess. I was mistaken in thinking this could not be done. We see again the interpenetration of theories of knowledge and inference.

Princeton University

BIBLIOGRAPHY

[1] Michael Clark, 'Knowledge and Grounds: A Comment on Mr. Gettier's Paper', *Analysis* **24** (1963) 46–8.
[2] Edmund L. Gettier, 'Is Justified True Belief Knowledge?', *Analysis* **23** (1963) 121–3.
[3] Alvin Goldman, 'A Causal Theory of Knowing', *Journal of Philosophy* **64** (1967) 357–72.
[4] A Phillips Griffiths (ed.), *Knowledge and Belief*, Oxford University Press, London, 1967.
[5] Gilbert Harman, 'How Belief is Based on Inference', *Journal of Philosophy* **61** (1964) 353–9.
[6] Gilbert Harman, 'The Inference to the Best Explanation', *Philosophical Review* **63** (1966) 241–7.
[7] Gilbert Harman, 'Unger on Knowledge', *Journal of Philosophy* **64** (1967) 390–5.
[8] Gilbert Harman, 'Detachment, Probability, and Maximum Likelihood', *Nous* **1** (1967) 401–11.
[9] Gilbert Harman, 'Knowledge, Inference, and Explanation', *American Philosophical Quarterly* **5** (1968) 164–73.
[10] Gilbert Harman, 'Enumerative Induction as Inference to the Best Explanation', *Journal of Philosophy* **65** (1968) 529–33.
[11] Keith Lehrer, 'Knowledge, Truth, and Evidence', *Analysis* **25** (1965) 168–75.

[12] Isaac Levi, *Gambling with Truth*, New York 1967.
[13] William W. Rozeboom, Jr., 'Why I Know So Much More Than You Do', *American Philosophical Quarterly* **4** (1967) 257–68.
[14] E. Sosa, 'The Analysis of "Knowledge that *p*"', *Analysis* **25** (1964) 1–8.
[15] J. T. Saunders and N. Champawat, 'Mr. Clark's Definition of 'Knowledge"', *Analysis* **25** (1964) 8–9.
[16] Brian Skyrms, 'The Explication of "X knows that *p*"', *Journal of Philosophy* **64** (1967) 373–89.
[17] Peter Unger, 'Experience and Factual Knowledge', *Journal of Philosophy* **64** (1967) 152–73.
[18] Peter Unger, 'An Analysis of Factual Knowledge', *Journal of Philosophy* **65** (1968) 157–70.

REFERENCES

* Many people have commented on earlier drafts of this paper. I am particularly grateful to Ronald De Sousa and Thomas Nagel.

1 Edmund L. Gettier [2], reprinted in A. Phillips Griffiths [4], 144–6.

2 Michael Clark [1]; Alvin Goldman [3]; Gilbert Harman [5]–[10]; Keith Lehrer [11]; William W. Rozeboom, Jr. [13]; E. Sosa [14]; Saunders and Champawat [15]; Brian Skyrms [16]; Peter Unger [17] and [18].

3 Cf. Rozeboom [13].

4 A speaker of English is much more inclined to say that Jones knows in the first case but not in the second than he is to say Jones knows in the second case but not in the first. We must account for that difference.

5 Goldman [3].

6 *Ibid.*, p. 368.

7 For some indication of how I would attempt to give a more precise statement of (c), see my [5]. Nothing in the present paper depends on its precise statement. It is perhaps worth noting that belief can be based on a great number of different inferences, only one of which need satisfy (a) and (b) for a man to have knowledge.

8 Harman [10].

9 Harman [8].

10 E.g. to get *X* from *X explains Y*.

11 In his Princeton Ph.D. thesis, Robert Nozick points out that deductive 'rules of inference' do not tell one what to believe. He and others have suggested that such rules might be construed as telling one what it is irrational to believe. That would be to accept the following principle:

(A) It is irrational to believe propositions that are jointly logically inconsistent.

At the Penn conference, (A) was endorsed by Marshall Swain, Keith Lehrer, and possibly Isaac Levi. Henry Kyburg objected to (A) on the grounds that one might be justified in believing each thing one believes, while also being justified in believing that at least one thing one believes is false. Nozick independently (and tentatively) raised the same objection.

Even more serious is the following objection to (A). It would not be irrational to believe propositions which are jointly logically inconsistent if one could not be expected to realize they are. At best (and only if we dismiss the Kyburg-Nozick point), one must modify (A) in the direction of (B):

(B) It is irrational to believe propositions that one knows (or ought to know) are jointly logically inconsistent.

But if that's the principle, why bring in logic? Why not simply say the following?

(C) It is irrational to believe propositions which one knows (or ought to know) cannot be true together.

Principle (B) is no more plausible than principle (C) which entails it; but (C) has nothing particularly to do with deductive logic. Logic can tell one that certain propositions cannot be true together; but so can mathematics, physics, common sense, etc.

[12] If h is some hypothesis and e is the total evidence, the conditional probability of h on e, written $p(h, e)$, is defined to be $p(h \& e)/p(e)$, where $p(h \& e)$ is the initial probability that h and e are both true and $p(e)$ is the initial probability of e.

[13] If e logically entails h, $p(h \& e) = p(e)$ in the standard probability calculus; so $p(h, e) = p(h \& e)/p(e) = 1$.

[14] E.g. Isaac Levi [12].

[15] Roughly speaking, a subjective probability represents a person's more or less rational 'degree of belief' in a proposition.

[16] Recommend by R. C. Jeffrey.

[17] See note 7 above.

[18] Harman [9], Section VIII, fn. 11, p. 163, which anticipates the present discussion was added in proof.

[19] The word 'special' is and is meant to be a fudge.

[20] *Op. cit.*

KEITH LEHRER

JUSTIFICATION, EXPLANATION, AND INDUCTION

A belief is justified by its explanatory role in a system of beliefs. Some beliefs are justified because of what they explain, and other beliefs are justified because they are explained, but every belief that is justified is so either because of what it explains or because of what explains it. These doctrines formulate a theory of justification which I shall label 'explanationism'. The theory has much support among contemporary philosophers, even though it is rarely stated this explicitly, and no general defense of it will be attempted here.[1] Instead, I shall assume that the theory is correct and then proceed to consider two problems that must be solved by a philosopher who is committed to explanationism. The first problem concerns the question of how those basic beliefs which provide what is to be explained as well as what is to do the explaining can be justified within this theory. The second problem concerns the question of when a hypothesis is inductively justified and how inductive justification can be specified in terms of explanation. I shall attempt to formulate these problems with greater precision below, and, having done so, propose solutions to them.

I

Before considering the two problems mentioned above, I shall make some basic remarks concerning the nature of explanation. It has been argued by many philosophers that there are a great variety of ways in which something may be explained, and these various forms of explanation may all be considered scientific. So much diversity makes it more difficult to provide any general analysis of explanation. Hence my remarks shall be restricted to presenting some considerations that are important for my subsequent argument.

One form of explanation that has received considerable philosophical attention is that which explains some fact by deducing the fact to be explained from some law or theory and certain antecedent conditions. The attempts to analyze this familiar form of explanation have met with

Swain (ed.), Induction, Acceptance, and Rational Belief.

greater difficulty and less success than one would initially expect.[2] There is no analysis of such deductive explanation to which I would subscribe. However, certain results are clear enough. For my purposes the most important of these is that a deductive explanation may contain the fact to be explained as an essential conjunctive component. Thus, it is no objection to an explanation that the very thing to be explained occurs as an essential conjunct in the statement that constitutes the explanation.

To see that this is so, suppose that E is a premiss that would suffice for both the deduction and the explanation of some fact F. Now anything that is logically equivalent to E, which has exactly the same logical content as E, must also suffice for the deduction and explanation of F. However, E is logically equivalent to a conjunctive statement, namely, the conjunction of F, which is a deductive consequence of E, and the statement $F \supset E$. It is obvious that E is equivalent to the conjunction of F and $F \supset E$, because from that conjunction we may deduce F and E, and from E we may deduce the conjunction.[3] Moreover, if E explains F, then $F \supset E$ will not explain F. Thus in order to explain F in terms of the conjunction of F and $F \supset E$, we need to employ F as a premiss to deduce E which explains F. So F is essential in the conjunction for the explanation of F; if F is dropped from the conjunction, the remaining conjunct will prove insufficient for the explanation of F.

A more concrete example may be more illuminating if more controversial. Suppose we wish to explain why a certain animal a is sterile, and this is explained by the fact that a is a mule and the law that all mules are sterile. From the conjunction of this fact and law we can deduce and explain that a is sterile. However, the statement 'a is a mule and all mules are sterile' is logically equivalent to the statement 'a is sterile and (a is sterile $\supset$ [a is a mule, and all mules are sterile])'. The second major conjunct of the latter statement 'a is sterile $\supset$ (a is a mule, and all mules are sterile)' is not a statement from which we may deduce 'a is sterile' and therefore is not sufficient to explain the fact that a is sterile. Indeed, it is only by using the first conjunct 'a is sterile' that we can derive the statement 'a is a mule, and all mules are sterile' which does explain the fact in question. Thus the statement 'a is sterile' is essential in the conjunction for the explanation of the fact that a is sterile. Therefore, an explanation may be satisfactory even though it contains a statement of the thing explained as an essential conjunct in the statement of the explanation.

The preceding remarks concern deductive explanation. The problems confronting the analysis of inductive explanation are equally perplexing, and again we lack a completely satisfactory analysis.[4] Inductive explanation is explanation by the probabilities. For example, suppose that I notice that a slot machine has paid off $80 on $100 played and wonder how this is to be explained. One explanation is that the machine is set so that the most probable payoff on a play of $100 is $80. This explanation is more satisfactory once we add the information that the payoff of $80 on $100 is not only the most probable, but is also more probable than 1/2. From this information it would be reasonable to expect what I have observed. Of course, one could inquire further and seek an explanation for the probabilities being what they are. But nevertheless, from the information concerning the probabilities of various possible payoffs, we may explain the payoff. The fact that $80 was paid on $100 played is inductively explained by the probabilities.

The preceding considerations suggest a connection between explanation and expectation, namely, that if some fact is explained by our knowledge, then it would be reasonable to expect such a fact on the basis of that knowledge. Thus, reasonable expectation is a necessary condition of explanation. It might seem plausible to suppose that reasonable expectation is sufficient for explanation, but this is not so. Anything that I know would be reasonable for me to expect on the basis of that knowledge, but not everything that I know is explained by my knowledge. Moreover, even when I cannot deduce from my knowledge what I reasonably expect, we have no guarantee of an explanation. Suppose I know that there is a mountain marked on a map I possess, and from my knowledge I expect, quite reasonably, that there is in fact a mountain in a certain location. I cannot deduce from my knowledge of the map that the mountain is so located, because it is possible that the map is defective. However, it is quite clear that my knowledge of the map does not explain the presence of the mountain in the location marked.[5] Here is nondeductive expectation which, though reasonable, leaves the expected fact unexplained. Thus, reasonable expectation is not sufficient but is necessary for explanation.

After a summary of these remarks on explanation, I shall proceed to the primary concerns of the paper. I have argued that an explanation of some fact may contain a statement of that fact as conjunct. This is demonstrable in the case of deduction, and I shall assume that it is true of explanation in general. Next I shall assume that a fact may be explained by the probabilities. Finally, I shall regard reasonable expectation of a fact on the basis of knowledge as a necessary condition for explanation of the fact by such knowledge.

II

We have said that according to explanationism every belief that is justified is so either because of what it explains or because of what explains it. However, such explanation presupposes that there are some beliefs that we assume to be true, otherwise there will be nothing to explain and nothing to be explained. In short, justification in terms of explanation will normally proceed against a background of beliefs that are already included within the body of knowledge. The problem is how such beliefs can be justified within the assumptions of explanationism.

Some philosophers have argued that there are beliefs which constitute an inviolable and irrefutable foundation for the justification of all other beliefs, but such a doctrine is alien to the doctrine of explanationism.[6] The reason is that we must allow that what is explained by certain beliefs at one time might not be explained by those beliefs at another, and, consequently, that what is justified at one time might not be justified at another and vice versa. Thus, explanationism leads to the conclusion that nothing is irrefutable or inviolable. To use a term from Peirce, explanationism is a fallibilistic theory.

A defender of explanationism, having found himself a bedfellow of pragmatism, may be tempted to argue that there are always beliefs that we embrace, indeed, many beliefs which we cannot help but have, and the existence of such beliefs may be considered the fallible but adequate starting point for explanation, and hence for justification. But this is a bad argument. Some beliefs that we have are unjustified and unreasonable, and the same is true of some beliefs that are completely compelling. Such beliefs, if countenanced, would be the starting point for little but confusion. No doubt some beliefs that exist are justified and do constitute a reasonable basis for justification of other beliefs. The problem, however, is to specify which of those beliefs are justified and to show how the explanatory role of such beliefs can account for their justification.

How can we show that some belief is justified either because of what it explains or because of what explains it without appealing to any other belief which it explains or which explains it? That is the crucial question, and the answer is not hard to find. To do so, let us consider the kind of belief that has been the most common candidate for the role of first member in the chain of justification, beliefs concerning perceptual experience. To take a concrete example, suppose I believe that I see a pen on my desk. This belief is, I contend, both justified and fallible. Perceptual beliefs of this sort are sometimes erroneous and yet they are also sometimes justified. But what explains this belief or what is explained by it which makes it justified?

The answer to the preceding question depends on recognizing that the existence of such a belief is a fact which itself demands explanation. If I believe that I see a pen on my desk, then the fact that I have this belief, that this belief exists, is something to explain. To the question, why does he believe that?, we must find an explanatory answer. And this shows how the belief is justified. For, one explanation of why I believe that I see a pen on my desk is simply that I do see a pen on my desk. The existence of this belief, like many other perceptual beliefs, is explained by the truth of the belief. If it is true that I see what I believe I see, this is often a clear and correct explanation of why I believe what I do. Thus, some perceptual beliefs, the most familiar ones, are justified because the existence of the belief is most simply explained by assuming that it is true. Moreover, as in the case of my belief that I see a pen on my desk, it would be difficult in many instances of perceptual belief to explain the existence of the belief if the belief were not assumed to be true. Why, after all, should I believe that I see a pen on my desk if I do not see one? The answer to this particular question would have to be both complex and fantastic.

I have argued that some perceptual beliefs are explained by their truth, and, consequently, that one is justified in such beliefs. However, the word 'some' is a problem. Clearly, not all perceptual beliefs are explained by the assumption that they are true, for some such beliefs are absurd and others are too sophisticated. The belief that I see pink rats is too absurd to be explained by the assumption that it is true, and the belief that I see an electron path is too sophisticated to be explained by an analogous assumption. The solution to the problem depends on two considerations. The first is that some perceptual beliefs may be explained best on the assumption that they are false; we may best explain my belief that I see pink rats on the assumption that the belief is false. Secondly, some beliefs, though perceptual, depend on rather complex background knowledge, and they may only be explained on the assumption that the person in question has such knowledge. Thus, my belief that I see an electron path depends on knowledge about the behavior and existence of atomic particles. It is, of course, controversial whether there are any perceptual beliefs that may be explained in complete independence of all background knowledge, and I would concede for the sake of argument that perhaps the best explanation of the existence of any perceptual belief will depend on the information that we have concerning the background knowledge of the subject. However, those perceptual beliefs about what we see immediately before our eyes in the most favorable conditions which do not depend on any esoteric or complicated knowledge, may be explained on the assumption they are true even if such explanation is neither complete nor the best that it is possible to produce. Hence, those perceptual beliefs whose existence may be explained without appeal to background knowledge of the subject and without assuming the falsity of the belief, are justified if the existence of such a belief may be explained by the assumption that the belief is true.

According to explanationism, a belief is justified either because of what it explains or because of what explains it. This means that if *S* is justified in believing *P*, either *P* explains some fact or *P* is a fact which is explained. That I see a pen on my desk explains the fact that I believe that I see a pen on my desk, and, hence, I am justified in believing that I see a pen on my desk. In the case of such a perceptual belief, my believing *P* is justified because of what *P* explains. The only peculiarity is that what *P* explains is my believing *P*. In this way, the belief is self-justifying.[7]

The preceding remarks illustrate the way in which an explanationist theory of justification can retain a basic connection with empiricism and at the same time reject the idea that there are any beliefs that constitute the irrefutable foundation of all justification. It is perceptual beliefs whose existence is explained by their truth, and this provides for the special position afforded such statements by empiricist theories of knowledge. General beliefs, such as my belief that all men are mortal, are not beliefs whose existence is explained by their truth. It does not at all explain why I believe that all men are mortal to say that it is true that all men are mortal, because a general fact about men does not explain my believing that truth in the way in which a truth about what I see may explain my believing that truth. Hence perceptual beliefs are justified by their very existence in a way in which other beliefs are not. However, such justification and explanation of perceptual beliefs is entirely fallible. If a better explanation for the existence of such beliefs turns on the assumption that the belief is false, then the existence may no longer be explained by the assumption of truth and consequently the belief is no longer justified.[8]

The theory of justification I have been formulating remains fragmentary. One would have to offer some analysis of what makes an explanation better than another, of what makes an explanation more complete than another, of what makes an explanation simpler than another, and so forth, in order to provide a thoroughly satisfactory theory based on explanationist presuppositions. However, having argued that it is possible to offer a justification of certain beliefs in terms of the explanatory role of such beliefs without appeal to any background of justified beliefs, I conclude that explanationism can account for the existence of a background of justified belief without abandoning the basic presuppositions of the doctrine.

Having shown how certain beliefs may be justified in terms of explanation without appeal to background knowledge, I shall now turn to the problem of specifying when a hypothesis that we believe is inductively justified by our background knowledge. The supposition that such knowledge exists is warranted by the preceding discussion. Since we may account for a background of justified beliefs within the theory of explanationism, we may reasonably suppose that some of these beliefs satisfy the additional conditions, such as that of being true, requisite to knowledge.[9]

III

Let us now turn to the problem of inductive justification. The question is – when is a hypothesis inductively justified by background knowledge? There are other ways to formulate the same question. The following questions are equivalent:

(1) Under what conditions is a hypothesis inductively justified by background knowledge?

(2) Under what conditions may a hypothesis be inductively inferred (induced) from background knowledge?

Given the theory of explanationism, these are also equivalent to

(3) Under what conditions is a hypothesis inductively explained by background knowledge?

In order to clarify the use of the term 'inductively' employed in these questions, it may be useful to contrast inductive and deductive inference.

The first and most basic difference between deductive inference and inductive inference is that the former is truth preserving while the latter is not.

Consider the following two principles:

(P1d) If *h* may be deductively inferred from *b*, then *h* is true if *b* is true.

(P1i) If *h* may be inductively inferred from *b*, then *h* is true if *b* is true.

Of these two principles, the former is true while the latter is false. Inductive inference is not generally truth preserving. This is not a defect of inductive inference but is rather a characteristic of it.

This characteristic led Hume to argue that there is no reasoning by which one can justify an inductive inference.[10] If it be demanded that such reasoning show that the conclusion of an inductive inference is true when the premisses are, then Hume is correct. But, truth is not the only desideratum of inference, another is explanation. Consequently, though we cannot prove that the conclusion of an inductive inference is true if the premisses are true, we can prove that the conclusion is explained if it is true. By proving that the conclusion is explained if true, we may justify the inference without proving that the conclusion is true.

That inductive inference is not truth preserving, should lead us to expect that it may lack other features of deductive inference that result from the truth preserving character of deduction. However, the lack of such features should not lead us to sceptical rejection of inductive inference, for such a policy is as unreasonable as Hume's. Consider the following two principles:

(p2d) If *b* is known to be true, and *h* is deductively inferred from *b*, then *h* is known to be true.

(p2i) If *b* is known to be true, and *h* is inductively inferred from *b*, then *h* is known to be true.

Since the former principle is true and the latter untrue, one cannot directly extend knowledge by induction as one can by deduction.[11] But this is no argument whatever against inductive inference. For, the falsity of (p2i) is a simple consequence of the fact that induction is not truth preserving.

It has also been argued that the requirement of total evidence is needed in the case of inductive inference but not in the case of deductive inference. The need for this requirement in the case of inductive inference is alleged to arise because additional evidence can undermine inductive inference but cannot undermine deductive inference.[12] This argument can be clarified by considering the following two principles:

(p3d) If *h* may be deductively inferred from *b*, then for any *i*, *h* may be deductively inferred from *b* and *i*.

(p3i) If *h* may be inductively inferred from *b*, then for any *i*, *h* may be inductively inferred from *b* and *i*.

It is again the case that the first principle is true and the second false. But this is also a consequence of the fact that deduction is truth preserving and induction is not. When *h* may be inductively (but not deductively) inferred from *b*, then *h* may be false even though *b* is true, and so we cannot inductively infer *h* from the conjunction of *b* and the denial of *h*. This contrasts with the case of deduction, where if we may deductively infer *h* from *b*, then we may also deductively infer *h* from *b* and the denial of *h*.

However, when either deduction or induction is employed for the purposes of justification, the addition of further statements to the premisses of the justificatory argument may alter the argument so that it justifies nothing. If we have a deductive argument intended to justify *h*, any justificatory value that the argument might have can be quickly eliminated if we add the denial of *h* to the premisses. For contradictory premisses justify nothing, though everything may be deduced from them.

The preceding remarks are intended to offer some clarification of the character of inductive inference and justification. However, the primary task of formulating some principle of inductive justification still remains. Such principles have proved very controversial for at least two reasons. First, it is often unclear what conditions such a rule should satisfy. The objectives of inductive inference have not always been clearly formulated. In the case of deductive logic, the objective is truth preservation. Inductive inference, we have noticed, is not generally truth preserving. But, if truth preservation is the goal of inductive inference, then, one might ask, what is the goal of inductive inference? The goal of inductive inference, I have argued, is to arrive at conclusions which, if true, are explained by our background knowledge.

Second, it has been objected that such rules of inductive inference are unnecessary. Once we know the probability of various hypotheses with respect to our background knowledge, we can employ such probabilities to arrive at a decision concerning any practical decision we have to make.[13] Thus, no inductive inference from our background knowledge to any hypothesis is required. The reply to this line of thought is the same as to the preceding. There is more to science and epistemology than the preservation of truth and practical decision making. There are other objectives of inquiry. One of these is explanation. Again, explanation is the objective of induction.

Thus, we shall approach the subject of inductive inference with the assumption that a hypothesis *h* is inductively (but not deductively) justified by background knowledge *b* only if *h* is explained by *b*. Two qualifications are necessary. First, though I regard explanation as the objective of inductive inference, I do not deny that the attainment of this objective may entail the attainment of others as well. I. Levi has proposed truth and the relief from agnosticism as objectives of inductive inference, and I would contend that the attainment of these objectives is entailed by the rule I shall consider.[14] Nevertheless, the primary goal is explanation, and the other goals are derivative.

Second, some philosophers have argued that inductive inference is inference to what explains rather than to what is explained.[15] Gilbert Harman has argued that such inference is inference to the best explanation.[16] Any apparent disagreement with this point of view would be more verbal than real. I agree that we are justified in believing some hypotheses because they explain our background knowledge, just as I maintain that we are justified in believing other hypotheses because they are explained by such knowledge. Indeed, we are justified in believing hypotheses precisely because they increase the explanatory coherence of our entire system of beliefs. Thus, my choice of the word 'inductive' to describe nondeductive inference from our background information to what is explained by that knowledge is not to be construed as excluding other forms of nondeductive inference. Following Peirce, I propose that we call nondeductive inference from our background knowledge to what explains it 'abductive' inference – rather than inductive inference. However, there is a large measure of verbal stipulation, I admit, in this usage.

IV

I shall now attempt to formulate some conditions concerning explanation to guide our search for an adequate rule of inductive inference. Letting the locution '$E(h, b)$' mean 'h is explained by b', the first principle, which I shall call the principle of the conservation of explanation, for short, the *conservation principle*, is as follows:

(E1) If $E(h, b)$ and $E(k, b)$, then $E(h, b \& k)$.

This principle asserts that if a hypothesis is explained by our background knowledge, then that hypothesis is also explained by our background knowledge expanded to include any other statement explained by it. The reason for accepting it is that if our knowledge explains something, the explanation cannot depend on the falsity of another hypothesis explained by the same knowledge. Anything that is explained by our knowledge must be explained even if every other hypothesis explained by the evidence turns out to be true. This is what the conservation principle requires.

If we accept the conservation principle, then we must accept the consequence that a statement may be explained by a conjunction of which it is a conjunct. But we have seen from the earlier discussion of explanation that the most straightforward cases of explanations are logically equivalent to an explanation which contains the thing to be explained as a conjunct. So this consequence is acceptable.

The next condition concerns the explanation of a set of hypotheses. It is as follows:

(E2) If S is a nonempty set of hypotheses h_j such that $E(h_j, b)$, then S is explained by b, that is, $E(S, b)$.

This is a condition that asserts no more than that the set of all those hypotheses that are explained by our background knowledge is itself explained. This condition is unlikely to require defense. However, the following principle of the conjunctivity of explanation, which I shall refer to as the *conjunction principle*, is more controversial:

(E3) If $E(S, b)$ and C is a conjunction of the members of a finite subset of S, then C is explained by b, that is, $E(C, b)$.[17]

From E2 and E3 it follows that

If $E(h, b)$ and $E(k, b)$, then $E(h \& k, b)$.

The latter principle needs clarification. It says that if one hypothesis is explained by our background knowledge and so is a second hypothesis, then the conjunction of those two hypotheses is also explained. Often when a person asks why two things both occurred, why the footprints of Jones were found outside the window and why his gun was found in the room, the person is implying that there is some connection between the two facts. However, this implication may be unwarranted, the explanation of the conjunction may consist of an explanation of each conjunct, and this may show that there was no connection between the two facts. For example, it might be explained that Jones left his footprints outside the window while capturing a butterfly and the gun was left inside by the murderer to incriminate Jones. Once each of the conjuncts of a conjunction has been explained by our background knowledge, then the conjunction has been explained. But this does not imply that there is any explanatory connection between the two conjuncts. In short, a conjunction may be explained even though neither conjunct has anything to do with the explanation of the other. The conjunction of two hypotheses does not presuppose that there is any connection between those hypotheses, and, consequently, the explanation of such a conjunction does not presuppose the explanation of such a connection.

The preceding conditions lead to the conclusion that a set of explained hypotheses must be logically consistent – no contradiction may be deduced from the set. It is also reasonable to assume that our background knowledge cannot explain a set of hypotheses logically inconsistent with that knowledge. Thus, we must add a principle of the consistency of explanation. I propose the following *consistency principle*:

(E4) Any nonempty set of hypotheses S, such that $E(S, b)$, is logically consistent with b, that is, the union of S and the unit set of b is logically consistent.

It is an immediate consequence of this condition that the set S is logically consistent, because a logically inconsistent set is not logically consistent with anything. This condition must be accepted to make the preceding conditions coherent. For, if we suppose that a set of hypotheses could be explained by our background knowledge and also be logically inconsistent, then it follows from the preceding principles that some background knowledge could explain the conjunction of a contradictory set of hypotheses. But that would mean that our background knowledge could explain a hypothesis which could not possibly be true. Thus, the conjunction principle (E3) requires that we add a consistency condition in order to avoid the conclusion that our knowledge explains contradictions. However, the conservation principle (E1) also supports this requirement, especially when taken together with (E3). Just as it seems unacceptable to suppose that a contradiction can be explained by our knowledge, so it is unacceptable to suppose that a contradiction explains some hypothesis, and, it is even more absurd to suppose that a contradiction might explain a contradiction.

The foregoing principles are the only ones I shall defend here. However, there are two additional principles which are of some interest. The first, a principle of the deductive closure of explanation, called the *deductive closure principle*, is

(E5*) If $E(S, b)$ and h is a logical consequence of S, then $E(h, b)$.

The second, which is a modified transitivity principle of explanation, which I shall call the *transitivity principle*, is

(E6*) If $E(h, b)$ and $E(k, h \& b)$, then $E(k, b)$.

These two principles are difficult to judge on intuitive grounds. The deductive closure principle says that anything that is a logical consequence of something explained by our background knowledge is itself explained by that knowledge. The transitivity principle says that if we conjoin hypotheses explained by our background knowledge to that knowledge and explain a hypothesis by that conjunction, then that hypothesis is explained by our background knowledge.

Principle (E5*) implies the conjunction principle (E3), but it is not implied by that principle. The conjunction conditions tell us that a certain logical consequence of a set of hypotheses explained by our background knowledge is explained by that knowledge, namely, the conjunction of those hypotheses. This leaves entirely open the question of whether other logical consequences of such a set are explained. There are reasons for thinking that some are not. For example, suppose that our background knowledge would explain h but h is false. Moreover, suppose k is true but that our background knowledge explains the denial of k. The disjunction, $(h \vee k)$, is true, and, given deductive closure, (E5*), we must conclude that this disjunction is explained by our background knowledge because h is explained by it. Since the truth of the disjunction is entirely a consequence of the truth of k, it seems odd to suppose that the disjunction is explained by knowledge that explains the denial of k. Though the disjunction, $(h \vee k)$ and the disjunction, $(h \vee -k)$, are both logical consequences of the explained hypothesis h, the latter but not the former is a consequence of $-k$, which is also explained. Consequently, the latter disjunction and not the former is explained by our background knowledge.

Principle (E6*) says that if we conjoin any hypothesis explained by background knowledge to that knowledge, then anything explained by the resultant conjunction is explained by the background knowledge alone. This principle is plausible when the explained hypotheses are deduced from our background knowledge, but when the inference is inductive it is rather doubtful. The reason is that when explanation is deductive, an explained hypothesis will be a deductive consequence of the background knowledge, and the conjunction of that hypothesis to the background knowledge will be logically equivalent to the original knowledge. Thus, in the deductive case, the content of the background knowledge will not be changed by conjoining explained hypotheses to the background knowledge. However, when the explained hypothesis is inductively inferred from the background knowledge, the conjunction of such a hypothesis to that knowledge will not be logically equivalent to the original knowledge. Instead, the content of the conjunction will be greater than the content of the original background knowledge, and, consequently, there is no reason to think what is explained by such a conjunction is explained by the background knowledge alone. Thus, (E6*) is dubious. Both (E6*) and (E5*) are suspect and have for that reason been marked with an asterisk.

V

Having formulated the preceding conditions of explanation, let us now return to the problem of formulating a principle of inductive inference. Taking explanationism as our underlying theory of justification, we shall attempt to formulate a principle of induction that is also a principle of explanation. More precisely, our inductive principle should yield the result that if a hypothesis is inductively inferred from our background knowledge, then that hypothesis is explained by such knowledge. Consequently, any inductive principle that conflicts with conditions (E1)–(E5) above must be rejected as unsatisfactory. To insure that such a conflict will not result, let us accept a set of principles pertaining to induction that correspond to the principles of explanation just formulated.

Letting the locution '$I(h, b)$' mean '*h* may be inductively inferred from *b*', and substituting this locution for the locution '$E(h, b)$' in principles (E1)–(E4), we obtain the following set of principles.

(I1) If $I(h, b)$ and $I(k, b)$, then $I(h, b \,\&\, k)$.

(I2) If *S* is a nonempty set of hypotheses h_j such that $I(h_j, b)$, then *S* may be inductively inferred from *b*, that is, $I(S, b)$.

(I3) If $I(S, b)$ and *C* is a conjunction of the members of a finite subset of *S*, then *S* may be inductively inferred from *b*, that is, $I(C, b)$.

(I4) Any nonempty set of hypotheses *S* such that $I(S, b)$ is logically consistent with *b*, that is, the union of *S* and the unit set of *b* is a logically consistent set.[18]

From (I2) and (I3) we obtain

If $I(h, b)$ and $I(k, b)$ then $I(h \,\&\, k, b)$.

We have also affirmed that induction implies explanation, and this implication may be formulated as follows:

(IE) If $I(h, b)$ and *h* is not a deductive consequence of *b*, then $E(h, b)$.

Any inductive principle which satisfies these five conditions will cohere with the four principles of explanation. Any rule that fails to satisfy these conditions on induction may yield results inconsistent with those explanatory principles. It would be possible to defend these results independently of the explanationist program just as it would be possible to defend that program itself. But no such defense will be attempted here. Instead I shall assume these conditions and ask what sort of principle can be found to satisfy them. Such an inquiry is of interest even aside from the program of explanationism, because an inductive principle that coheres with these principles, and only such, constitutes a principle of inductive explanation.

Before proceeding to the problem of finding a rule that satisfies the conditions listed above, one further condition should be added which has been implicit in the preceding discussion. We have assumed that explanation and induction are unaffected by the substitution of logically equivalent sentences. This principle may be formulated as follows:

(L) If h is logically equivalent to h' and b is logically equivalent to b', then $\mathrm{E}(h, b)$ if and only if $\mathrm{E}(h', b')$ and $\mathrm{I}(h, b)$ if and only if $\mathrm{I}(h', b')$.

Moreover, it should be noted that the following two conditions are not required:

(I5*) If $\mathrm{I}(S, b)$ and h is a logical consequence of S, then $\mathrm{I}(h, b)$.
(I6*) If $\mathrm{I}(h, b)$ and $\mathrm{I}(k, h \mathbin{\&} b)$, then $\mathrm{I}(k, b)$.

We have already noted some reasons for doubting such conditions when we considered (E5*) and (E6*). Since condition (I5*) has been advocated and disputed, the rejection of this condition is of some importance.

VI

With these considerations, let us examine the problem of finding some inductive principle that satisfies conditions (I1)–(I4) as well as (IE) and (L). Inductive inference has traditionally been conceived of in terms of probability relations. Thus it is important to have some characterization of probability to serve as the basis of further discussion. We first consider a finite set, or partition, of hypotheses $k_1, k_2, \ldots, k_n$ and background knowledge b such that b is consistent with each k_j and implies that exactly one of the n k_js is true. Now consider the language L consisting of the preceding partition and truth functional combinations of members of the partition. For any such language, letting '$\mathrm{P}(h, s)$' mean 'the probability of h on s', and assuming that the statement to the right of the comma in the probability expression is logically consistent, we shall adopt the following calculus of probability:

(P1) $0 \leqq P(h, b) \leqq 1$.
(P2) If $\vdash \ulcorner b = b' \urcorner$ and $\vdash \ulcorner h = h' \urcorner$, then $P(h, b) = P(h', b')$.
(P3) $P(h, b) = 1$ if and only if $b \vdash h$.
(P4) If $b \vdash \ulcorner -(h \,\&\, k) \urcorner$, then $P(h \vee k, b) = P(h, b) + P(k, b)$.
(P5) $P(h \,\&\, k, b) = P(h, b) \times P(k, h \,\&\, b)$.

In the subsequent discussion it will be assumed that all inductive inference as well as probability assignments are based on a partition adequate to express the relevant alternatives under consideration. No further analysis of the concept of probability will be offered. I shall leave open the question of how the probability function might be further specified with the conviction that subsequent discussion will help to clarify that issue.

In seeking to formulate an inductive principle in terms of the probability relations that obtain among the sentences of a language, it will be useful to first consider a simple principle which violates the conditions on induction so that we may discover its defects. It is plausible to suppose that there is some degree of probability less than unity which suffices for inductive inference. Thus, where m and n are positive integers, m less than n, consider the following principle:

(R1) $I(h, b)$ if and only if $P(h, b)$ is at least m/n.

This principle conflicts with the inductive principles we have accepted.

To see why this is so, let us consider the case of a lottery in which we can assign probabilities in a reasonable manner and consider the results. Suppose that we have a lottery with n tickets and that our background information 'b' implies that exactly one ticket will be picked in a fair draw. Let h_1 say that the number 1 ticket will be picked, h_2 the number 2 ticket, and so forth. The set of hypotheses h_1, h_2, and so forth to h_n is a partition relative to the background knowledge. In other words, if b is background knowledge, then b implies that the set of hypotheses is exhaustive, one must be true, and exclusive in pairs, no two may be true. The negations of the hypotheses belonging to the partition, that is, the set containing $-h_1$, $-h_2$, and so forth to $-h_n$ generate the conflict in question. This set of hypotheses has the characteristic that for each negative hypothesis $-h_j$, the probability of that negative hypothesis on the background knowledge is at least m/n. The reason is that for each hypothesis,

h_j, $P(h_j, b) = 1/n$, and, hence $P(-h_j, b) = 1 - 1/n$. Of course, $1 - 1/n$ cannot be less than m/n if both m and n are positive integers and m is less than n. Since $P(-h_j, b)$ is at least m/n, given (R1) it follows that $I(-h_1, b)$, $I(-h_2, b)$ and so forth to $I(-h_n, b)$.[19]

The set S of hypotheses $-h_1$, $-h_2$, and so forth to $-h_n$ may be inductively inferred from b. From

(I2) If S is a nonempty set of hypotheses h_j such that $I(h_j, b)$, then $I(S, b)$.

and

(I3) If $I(S, b)$ and C is a conjunction of the members of a finite subset of S, then $I(C, b)$.

it follows that the conjunction of the members of S may also be inductively inferred from b. However, this conjunction has a probability of 0 on b, because b says exactly one ticket will be picked. Since (R1) says that a hypothesis may be inductively inferred from b only if the probability of it on b is at least m/n, (R1) fails to satisfy (I2) and (I3). Similarly, since the set S is logically inconsistent with b, (R1) fails to satisfy (I2) and (I4).

Both conditions (I3) and (I4), when combined with (I2), have been rejected by some philosophers. However, the arguments for such rejection are inapplicable in a context where one is attempting to formulate an inductive rule that is also a principle of inductive explanation. For, as we have seen above, our background knowledge explains the conjunction of any finite set of hypotheses explained by that knowledge, and no knowledge can explain a set of hypotheses inconsistent with that knowledge. Moreover, anything explained by our background knowledge will also be explained by any conjunction consisting of such knowledge and a hypothesis explained by it. This is asserted by the conservation principle for explanation which necessitates a conservation principle for induction,

(I1) If $I(h, b)$ and $I(k, b)$, then $I(h, b \& k)$.

However, this condition is not satisfied by (R1).

That (R1) does not satisfy (I1) is shown by the fact that the hypotheses $-h_1$, $-h_2$ and so forth are negatively relevant to each other on evidence b. Thus suppose that $m = n - 1$, which is the largest value it could have. By (R1) it follows that $I(-h_1, b)$ and $I(-h_2, b)$ because the probability of each of these hypotheses on b is m/n. But the probability of $-h_1$ on the conjunction of b and $-h_2$ is less than m/n. Hence it is not the case that $\mathrm{I}(-h_1, b \& -h_2)$. A similar argument can be constructed for lower values of m, provided $1/2 \leqq m/n < 1$. Suppose $m = n - c$, and let c be any positive integer such that the foregoing proviso is met. Consider the conjunction of the hypotheses $-h_1$, $-h_2$ and so forth to $-h_c$. This conjunction has a probability of m/n on b. Call this hypothesis k_1. From (R1) it follows that $\mathrm{I}(k_1, b)$. Secondly, consider the conjunction of the hypotheses $-h_{c+1}$, $-h_{c+2}$, and so forth to $-h_{c+c}$. Call this hypothesis k_2. The probability of k_2 on b is again equal to m/n, and by (R1) it follows that $\mathrm{I}(k_2, b)$. But it is not the case that $\mathrm{I}(k_1, b \& k_2)$ or that $\mathrm{I}(k_2, b \& k_1)$. Thus, for any value of m/n less than unity and not less than 1/2, the only values of any interest to us, condition (I1) is not satisfied. On the basis of (I1) and (E1), we must reject (R1).

There is an important moral to this story. Some philosophers, for example H. E. Kyburg, Jr., have argued that it would not be unreasonable to accept a set of hypotheses that are inconsistent with the background information from which they were inferred. To avoid some untoward consequences of this view, Kyburg abandons a condition that would enable us to conclude the acceptance of the conjunction of the inconsistent set of hypotheses.[20] We can accept inconsistent results, according to this view, provided we do not put two and two together and deduce a contradiction. Since this author construes his task as being one of formulating conditions of rationality, he maintains an extreme position. Surely rationality requires that we do not contradict in our conclusions what we accept in our premises. However, the issue need not be left at this level of radical disagreement. By considering principles of explanation, we have seen that condition (I1) must be accepted. But principles that do not satisfy a consistency requirement may be expected not to satisfy (I1) either. For, a set of hypotheses inductively inferred from our background knowledge which is inconsistent contains hypotheses that are negatively relevant to each other. Consequently, when one of these hypotheses are added to the background knowledge, those hypotheses to which they are negatively relevant lose some probability and may no longer be inductively inferred.

How may we formulate a more satisfactory principle? The key to the problem is the set of hypotheses $-h_1$, $-h_2$ and so forth. This set of hypotheses may fairly be regarded as competing with each other. It is not initially apparent that they are competitive because they are consistent in pairs. Nonetheless, they are competitive because they are inconsistent as a set relative to the background knowledge. Moreover, if any one of these hypotheses is deleted from the set, the set is no longer inconsistent relative to that knowledge. Thus the set is not merely inconsistent but *minimally* inconsistent relative to that knowledge – all the hypotheses are needed to obtain the inconsistency. This naturally suggests the requirement that an inductively inferred hypothesis should compete favorably with the other members of a set to which it belongs, when that set is a minimally inconsistent set relative to the background knowledge. And a hypothesis competes favorably with a hypothesis if it is more probable on such knowledge. This suggestion will enable us to formulate a rule of induction that satisfies all of the conditions formulated above.

VII

As a first step toward formulating such a rule, let us define the notion of a minimally inconsistent set, introducing the locution '*MIS*' to refer to such a set.

(DMIS) A set *S* is *MIS* if and only if *S* is inconsistent and every proper subset of *S* is consistent.

This notion of a set that is *MIS* is equivalent to a more familiar notion. Sometimes we speak of the consistent premises of a deductive argument as being essential for the deduction. The consistent premises of an argument are essential for the deduction of the conclusion if and only if the set of statements consisting of the premises of the argument and the denial of the conclusion is *MIS*. In addition to the notion of a set that is minimally inconsistent, we require the notion of a set that is minimally inconsistent relative to the background knowledge. Letting 'MIS_b' mean 'minimally inconsistent relative to *b*', we define the former as follows:

($DMIS_b$) A set *S* is MIS_b if and only if a set consisting of *b* and the members of *S* (and only that) is inconsistent, and every set consisting of *b* and the members of a proper subset of *S* (and only that) is consistent.

Having defined this locution we may proceed to the formulation of an inductive rule.

Our principle of induction is to say that a hypothesis may be inductively inferred from background knowledge just in case it competes favorably, that is, is more probable than those hypotheses with which it competes. The hypotheses with which a hypothesis competes are those that are members of the minimally inconsistent sets (relative to the background knowledge) which contain the hypothesis in question. The principle of induction is as follows:

(IR) $I(h, b)$ if and only if, (i) for any S, if h is a member of S and k is a member of S, $k \neq h$, and S is MIS_b, then $P(h, b)$ is greater than $P(k, b)$, and (ii) h is consistent with b.

Thus, for a hypothesis to be inductively inferred it must be more probable than other hypotheses belonging to those minimally inconsistent sets relative to the background knowledge which contain the hypothesis. All and only such hypotheses may be inductively inferred from the background knowledge.

It is important to note that IR requires for its application the specification of some partition of hypotheses $h_1, h_2, \ldots, h_n$, to which probabilities are assigned on the background knowledge. Moreover, once specified, the partition serves as a basis for the assignment of probabilities and inductive inference from background knowledge. I. Levi has argued that the specification of different partitions will yield different inductive results, and this result also holds for the application of (IR).[21] Hence, the problem of specifying a partition for the application of (IR) is not trivial. I cannot deal in detail with the problem of specification, but I can offer some relevant considerations. First, the partitions should be specified so as to explain as much as possible on the background knowledge. Second, the partitions should not contain arbitrary or irrelevant alternatives. For example, if we know a die has been tossed, when inferring the outcome we should not specify the partition: 'The one face turns up', 'the two face turns up', and 'some other face turns up', for the latter is arbitrary. Moreover, we should not specify the partition of five hypotheses each saying one of the first five faces turns up and the two hypotheses 'The six face turns up and is red', and 'The six face turns up and is not red', for these two hypotheses are irrelevant.[22] Finally, the set of partitions specified should yield consistent results so that we do not contradict ourselves. With these considerations in mind, let us consider the features of (IR).

It is clear that we do not obtain the results that followed from (R1) when we adopt (IR). In the lottery situation, none of the hypotheses $-h_1$, $-h_2$, and so forth may be inductively inferred from the background information according to (IR), because they constitute an equally probable set that is minimally inconsistent relative to the background information. Moreover, it is demonstrable that the rule satisfies principles (I1)–(I4) and does not satisfy conditions (I5*) and (I6*). We shall prove these results in the next section.

VIII

The conditions (I1)–(I4) are either derivable from (IR) or the consequences of those conditions concerning the inductive relations between statements are derivable from (IR). We may derive the Conservation Principle

(I1) If $I(h, b)$ and $I(k, b)$, then $I(h, b \,\&\, k)$.

from (IR).

PROOF: See Appendix, Proof 4.

Next consider condition

(I2) If S is a nonempty set of hypotheses h_j such that $I(h_j, b)$, then $I(S, b)$.

This condition is not derivable from (IR) because it asserts nothing about the inductive relations between any pair of statements. For this reason, it cannot by itself yield any consequences inconsistent with (IR). However, when (I2) is taken in conjunction with conditions (I3) and (I4), we do obtain consequences regarding the inductive relations between statements, and, these consequences must be shown to be derivable from (IR).

We noted earlier that from the Conjunction Principle

(I3) If $I(S, b)$ and C is a conjunction of the members of a finite subset of S, then $I(C, b)$.

and (I2) it follows that

If $I(h, b)$ and $I(k, b)$, then $I(h \,\&\, k, b)$.

The foregoing is the strongest consequence of (I2) and (I3) concerning the inductive relations between statements, and, consequently, once it is shown to be derivable from (IR), it follows that the conjunction of (I2) and (I3) is satisfied by (IR). We may prove that the statement in question is derivable from (IR).

PROOF: See Appendix, Proof 3.

Next we may prove that the Consistency Principle

(I4) Any nonempty set S such that $I(S, b)$ is consistent with b.

when taken in conjunction with (I2) is satisfied by (IR). These two conditions yield the consequence.

If S is a nonempty set of hypotheses h_j such that $I(h_j, b)$, then S is consistent with b.

Once we derive this consequence from (IR), it is obvious that the conjunction of (I2) and (I4) are satisfied by (IR).

PROOF: See Appendix, Proof 5.

IX

It was mentioned earlier that the Deductive Closure Principle is not required. We can also prove that (IR) fails to satisfy this principle in conjunction with (I2). The principle is as follows:

(I5*) If (S, b) and h is a logical consequence of S, then $I(h, b)$.

PROOF: Suppose that we have specified a partition consisting of h_1, h_2, and h_3 such that $P(h_1, b)=5/8$, $P(h_2, b)=2/8$, and $P(h_3, b)=1/8$. Then $I(h_1, b)$. Moreover, the set consisting of $(h_1 \vee h_3)$, $(h_1 \vee -h_3)$ and $-h_1$ is MIS_b. Though the two disjunctions are each consequences of h_1, they both are members of a set that is MIS_b. Hence it cannot be that both $I(h_1 \vee h_3, b)$ and $I(h_1 \vee -h_3, b)$.

Thus the basic reason that the Deductive Closure Principle is not satisfied by (IR) is that according to (IR) logical consequences of an hypothesis may be competitive and hence not all may be inferred.

The Modified Transitivity Principle, which was rejected earlier, is the following:

(I6*) If $I(h, b)$ and $I(k, h \,\&\, b)$, then $I(k, b)$.

This principle in conjunction with (I2) is not satisfied by (IR).

PROOF: Suppose that we have a specified partition consisting of h_1, h_2, and h_3 such that $P(h_1, b)=5/11$, $P(h_2, b)=4/11$, and $P(h_3, b)=2/11$. On this supposition, $I(-h_3, b)$, but it is not the case that $I(h_1, b)$. The latter is so because $P(-h_1, b)$ is greater than $P(h_1, b)$. However, $P(h_1, -h_3 \,\&\, b) = 5/9$, and, moreover, $I(h_1, -h_3 \,\&\, b)$.

Thus, if a hypothesis that is inferred from the background knowledge is conjoined to such knowledge, the set of alternatives may be reduced and the probabilities altered in such a way that a hypothesis may be inductively inferred that could not be inferred from the original knowledge.

Though I am willing to reject (I5*) and (I6*), it is interesting to notice a modification of (IR) that would permit the acceptance of the first of these conditions. In formulating (IR), we made a pair of conditions necessary and sufficient for the inductive inference of a hypothesis from background knowledge. Suppose, however, that we amend (IR) so that these two conditions are only asserted to be sufficient for such inference as follows:

(IRS) $I(h, b)$ if (i) for any S, if h is a member of S, and k is a member of S, $k \neq h$, and S is MIS_b, then $P(h, b)$ is greater than $P(k, b)$, and (ii) h is consistent with b.

We might then add a condition to the effect that $I(h, b)$ only if this is a consequence of (IRS) and (I1)–(I5*). Rule (IRS) is logically consistent with the conjunction of (I5*) and (I2), but as we have seen, the inductive relations that follow from this conjunction are not derivable from (IR) and hence are not derivable from (IRS). That the conjunction is consistent with (IRS) follows from the foregoing proof that the set of hypotheses h_j such that $I(h_j, b)$ is consistent with b.

A more interesting consequence of (IRS) would be the consistency of this principle with (I6*). I have no proof for this consistency. However, if principle (I6*) were consistent with (IRS), it would be a genuinely productive addition. Some inductive relations that are a consequence of it are not a consequence of (IRS). The latter is obvious from the proof that (I6*) is inconsistent with (IR). Moreover, if we were to add (I6*), then we would obtain (I5*) as a consequence. Having argued against the acceptance of (I6*), I shall not explore further the consequences of replacing (IR) with (IRS) and adding (I6*) except to note that one consequence of so doing would be to permit the inductive inference of hypotheses having a probability of less than 1/2 on background knowledge. Those who consider this an undesirable result may find this consequence a reason for adopting a more restrictive rule such as (IR).

The next problem is to show that (IR) is a principle of explanation. A hypothesis that is inductively inferred by (IR) from our background knowledge is one that we would reasonably expect to be true given such knowledge. If a hypothesis is more probable than any other hypothesis belonging to any minimally inconsistent set (relative to the background knowledge) containing the hypothesis, then it is only reasonable to expect such a hypothesis to be true. This is more apparent when one realizes that (IR) is logically equivalent to the following rule:

(IR_1) $I(h, b)$ if and only if (i) for any k, if k is not a logical consequence of the conjunction of h and b, then $P(h, b)$ is greater than $P(k, b)$; and (ii) h is consistent with b.

PROOF: See Appendix, Proof 6.

Given the equivalence of these two rules, we may conclude that a hypothesis inductively inferred from background knowledge is more probable than any other hypothesis except its logical consequences. It is reasonable to expect such a hypothesis to be true on the basis of that knowledge, and, if true, the hypothesis is explained by the knowledge. For, if h is true as well as inductively inferred by (IR), then, as is obvious from (IR_1), h is more probable on the background knowledge than any false hypothesis. This consequence is crucial from the standpoint of explanation, and (IR) is superior to inductive rules that do not yield this result. Suppose for example that instead of (IR) we adopt a rule that permits us to inductively infer that a ticket, say ticket number 10, will not win a certain skewed lottery because it is highly improbable that the ticket will be picked, even though there is another ticket, ticket number 11, which is even more unlikely to be picked. In the event that the number 11 ticket is picked, it is true that ticket 10 is not picked, but this is hardly explained by the probabilities. It was even more probable that ticket number 11 would not be picked than that ticket number 10 would not be, and ticket number 11 *was* picked. Hence, the fact that it was so improbable that the number 10 ticket would be picked fails to prove that the hypothesis 'the number 10 ticket will not be picked' is explained by our background knowledge when it turns out to be true. The reason is that some hypothesis that is even more probable on our background knowledge turned out to be false. (IR) rules out such a possibility for any true hypothesis that is inductively inferred from our background knowledge, and, by so doing, becomes effective as a principle of inductive explanation.

To support the contention that (IR) is a principle of inductive explanation, I shall consider a series of lotteries which reveal the implications of the rule. The cases involved are ones in which a partition specified for the application of the rule consists of the set of hypotheses h_1, h_2, and so forth to h_n, which assert that the number one ticket wins the lottery, the number two ticket wins, and so forth. Again the background knowledge asserts that one and only one ticket wins. The probabilities assigned to these hypotheses will be different in each case.

Case One: The lottery is fair. We have noted that since all the hypotheses of the partition are equally probable, none may be inferred from the background knowledge by (IR). Indeed, the only hypotheses that may be inferred are those deducible from that knowledge. This is appropriate, because no outcome of the lottery is inductively explained by the background knowledge.

Case Two: The number one ticket in the lottery is sticky and is less likely to be drawn than the others, but all the others are equally likely to be drawn. In this case, $-h_1$ may be inductively inferred from the background knowledge by (IR), but none of the hypotheses $-h_j$, $(j>1)$ may be so inferred. And again this is appropriate. It is only $-h_1$ that is explained by the background knowledge because it was the most probable. For any h_j, $(j>1)$, the probability of $-h_j$ is no greater than the probability of any other $-h_j$ and is less than the probability of $-h_1$. Consequently, such a hypothesis, even if true, is no more probable than another such hypothesis which is false, and, if h_1 is true, then it is less probable than the false hypothesis $-h_1$. In short, the hypotheses, $-h_j$, $(j>1)$ are not even on a par with those hypotheses in the fair lottery, in that, though they are equally probable, they are less probable than in the fair lottery. Thus are such hypotheses unexplained. On the other hand, if $-h_1$ is true, then there is no false hypothesis that is as probable, and thus it *is* explained by our knowledge.

Case Three: Ticket h_n is placed in a privileged position in the lottery so that the probability of h_n is greater than 1/2. In this case, h_n is the only member of the partition which may be inductively inferred from the background knowledge by (IR). If this hypothesis is false, then the true hypothesis h_j ($j<n$) is not explained because it was less probable than the false hypothesis h_n. On the other hand, if h_n is true, then since it was more probable than any h_j, it is also explained.

Proof that the foregoing results are obtained by (IR): See Appendix, Proofs 7–10.

An examination of these three cases and the proofs of the results reveals certain fundamental features of the application of (IR). First, a hypothesis may be inductively inferred from background knowledge only if it has a probability greater than 1/2 on the background knowledge. Second, though many hypotheses having a probability of 1/2 may not be inductively inferred, any hypothesis that is a member of the specified partition may be inductively inferred if it has a probability greater than 1/2 on the background knowledge. Third, a disjunction of the members of such a partition may be inductively inferred from the background knowledge if and only if such a disjunction is more probable than any other disjunction of such members which is not a logical consequence of it together with the background knowledge. The preceding characterization of the results of applying (IR) is complete, because any hypothesis consistent with the background knowledge which is a truth functional combination of members of the partition is logically equivalent to some disjunction of members of that partition. I conclude from this characterization that hypotheses inductively inferred from background knowledge by (IR) are ones it is reasonable to expect to be true on such knowledge and are explained by it. Thus, (IR) satisfies the objectives of an explanationist theory of justification.

X

We began by assuming an explanationist theory of justification, that every justified belief is justified either because of what it explains or because of what is explained by it. I have shown that it is possible to explain how certain beliefs, perceptual beliefs in particular, may be justified in terms of explanation without assuming any antecedent knowledge. Secondly, I have contended that it is possible to explain how beliefs based on inductive inference may be justified within the theory of explanationism and have provided an inductive rule to sustain my conclusion. These results support the theory of explanationism. Moreover, since my contentions are themselves explanatory, we may, by appeal to explanationism, conclude that they are justified. Thus, explanationism supports my results.

APPENDIX

A. *Calculus of Probability*

Consider a partition of hypotheses (a set of hypotheses that are disjoint in pairs and exclusive as a set) and a language L consisting of the members of the partition and truth functional combinations of those members. Assuming that h, k, and b are in L and that all the sentences occurring to the right of the comma in a probability expression are consistent, the following are axioms of probability:

(P1) $0 \leqq \mathrm{P}(h, b) \leqq 1$.

(P2) If $\vdash \ulcorner b \equiv b' \urcorner$ and $\vdash \ulcorner h \equiv h' \urcorner$, then $\mathrm{P}(h, b) = \mathrm{P}(h', b')$.

(P3) $\mathrm{P}(h, b) = 1$ if and only if $b \vdash h$.

(P4) If $b \vdash \ulcorner -(h \& k) \urcorner$, then $\mathrm{P}(h \vee k, b) = \mathrm{P}(h, b) + \mathrm{P}(k, b)$.

(P5) $\mathrm{P}(h \& k, b) = \mathrm{P}(h, b) \times \mathrm{P}(k, h \& b)$.

The following are theorems:

(PT1) $\mathrm{P}(h, b) + \mathrm{P}(-h, b) = 1$.

(PT2) If $h \vdash h'$, then $\mathrm{P}(h', b) \geqq \mathrm{P}(h, b)$.

B. *Rule of Induction*

Consider the following definition of a minimally inconsistent set relative to evidence:

(DMIS$_e$) A set S is MIS_b if and only if (i) $S \cup \{b\} \vdash \ulcorner p \& -p \urcorner$ and (ii) there is no set S' such that $S' \subset S$ and $S' \cup \{b\} \vdash \ulcorner p \& -p \urcorner$.

We may then formulate our inductive rule. Assuming that b, h, and k are in L, the rule is as follows:

(IR) $\mathrm{I}(h, b)$ if and only if (i) for any S, if $k \in S$, $h \in S$, $h \neq k$, and S is MIS_b, then $\mathrm{P}(h, b) > \mathrm{P}(k, b)$ and (ii) and it is not the case that $b \vdash \ulcorner -h \urcorner$.

C. *Logical Theorems*

(LT1) If $h \in S$ and $\ulcorner b \& k \urcorner \vdash h$, then S is not $MIS_{b \& k}$.

(LT2) If neither $b \vdash \ulcorner - k \urcorner$ nor $b \vdash \ulcorner - h \urcorner$ and neither $\ulcorner b \& h \urcorner \vdash k$ nor $\ulcorner b \& k \urcorner \vdash h$, then either $\{h, k\}$ is MIS_b or $\{h, k, \ulcorner - (h \& k) \urcorner\}$ is MIS_b.

(LT3) If $\vdash \ulcorner b \equiv b' \urcorner$ and $b \vdash \ulcorner h \equiv h' \urcorner$, then $h \in S$ and S is MIS_b if and only if $(S \cap \overline{\{h\}}) \cup \{h'\}$ is $MIS_{b'}$.

(LT4) If $\ulcorner b \& h \urcorner \vdash k$, then, if $j \in S$, $h \in S$, $j \neq h$ and S is $MIS_{b \& k}$, then S is MIS_b.

(LT5) If $k \in S$, $h \in S$, $h \neq k$, and S is MIS_b, then it is not the case that $\ulcorner b \& h \urcorner \vdash k$, or that $\ulcorner b \& k \urcorner \vdash h$.

(LT6) For any S in a language L consisting of the members of a partition, $h_1, h_2, \ldots, h_n$, and the truth functional combinations of such members, if it is not the case that $\vdash \ulcorner - S \urcorner$, then $\vdash \ulcorner S \equiv D \urcorner$ where D is either an h_j or a disjunction of all those h_js (each h_j occurring only once in the disjunction) such that it is not the case that $S \vdash \ulcorner - h_j \urcorner$.

D. *Proofs of Inductive Relations*

In the following proofs it is assumed that the sentences related in '$I(h, b)$' are sentences belonging to some language L specified as in A.

PROOF 1: If $\vdash \ulcorner b \equiv b' \urcorner$ and $\vdash \ulcorner h \equiv h' \urcorner$, then $I(h, b)$ if and only if $I(h', b')$. Assume $\vdash \ulcorner b \equiv b' \urcorner$ and $\vdash \ulcorner h \equiv h' \urcorner$. If $h \in S$ and S in MIS_b, then $(S \cap \overline{\{h\}}) \cup \cup \{h'\}$ is MIS_b by (LT3). Similarly, if $h \in S$ and S is MIS_b, then $(S \cap \overline{\{h'\}}) \cup \{h\}$ is MIS_b. Thus, for any k, there is a set S such that $k \in S$, $h \in S$, $h \neq k$ and S is MIS_b if and only if there is a set S such that $k \in S$, $h' \in S$, $h' \neq k$, and S is $MIS_{b'}$. Moreover, for any k, $P(k, b) = P(k, b')$ by (P2) just as $P(h', b') = P(h, b)$ by (P2). Thus, $P(k, b) < P(h, b)$ if and only if $P(k, b) < P(h', b')$. Therefore, $I(h, b)$ if and only if $I(h', b')$.

PROOF 2: If $I(h, b)$ and $I(k, b)$, then either $\ulcorner b \& k \urcorner \vdash h$ or $\ulcorner b \& h \urcorner \vdash k$. Assume $I(h, b)$ and $I(k, b)$. Suppose neither $\ulcorner b \& k \urcorner \vdash h$ nor $\ulcorner b \& h \urcorner \vdash k$. Since it is not the case that $b \vdash \ulcorner - k \urcorner$ or that $b \vdash \ulcorner - h \urcorner$ by (IR), the set $\{h, k\}$ is MIS_b or $\{h, k, \ulcorner - (h \& k) \urcorner\}$ is MIS_b by (LT2). It follows from our initial assumption and the preceding statement that $P(h, b) > P(k, b)$ and $P(k, b) > P(h, b)$. It is impossible that $P(h, b) > P(k, b)$ and $P(k, b) > P(h, b)$. Hence either $\ulcorner b \& k \urcorner \vdash h$ or $\ulcorner b \& h \urcorner \vdash k$.

PROOF 3: If $I(h, b)$ and $I(k, b)$, then $I(h \& k, b)$. Assume $I(h, b)$ and $I(k, b)$. Either $\ulcorner b \& k \urcorner \vdash h$ or $\ulcorner b \& h \urcorner \vdash k$ by Proof 2. Suppose $\ulcorner b \& k \urcorner \vdash h$.

In that case, if S is MIS_b and $k \in S$, then $(S \cap \overline{\{k\}}) \cup \{\ulcorner h \& k \urcorner\}$ is MIS_b by (LT3), and, by the same theorem, if $\ulcorner h \& k \urcorner \vdash S$ and S is MIS_b, then $(S \cap \overline{\{\ulcorner h \& k \urcorner\}}) \cup \{k\}$ is MIS_b. Hence, for any j, some S is such that $k \in S$, $j \in S$ and S is MIS_b if and only if some S is such that $\ulcorner h \& k \urcorner \vdash S$, $j \in S$ and S is MIS_b. Moreover, $\mathrm{P}(k, b) = \mathrm{P}(h \& k, b)$ because $\mathrm{P}(h \& k, b) = = \mathrm{P}(k, b) \times \mathrm{P}(h, k \& b)$ by (P5) and $\mathrm{P}(h, k \& b) = 1$ by (P3). Hence $\mathrm{I}(k, b)$ if and only if $\mathrm{I}(h \& k, b)$. From the supposition that $\ulcorner b \& k \urcorner \vdash h$ we have proven that $\mathrm{I}(h \& k, b)$, and an exactly similar proof that $\mathrm{I}(h \& k, b)$ may be given on the supposition that $\ulcorner b \& h \urcorner \vdash k$. Therefore, $\mathrm{I}(h \& k, b)$.

PROOF 4: If $\mathrm{I}(h, b)$ and $\mathrm{I}(k, b)$, then $\mathrm{I}(h, k \& b)$. Assume $\mathrm{I}(h, b)$ and $\mathrm{I}(k, b)$. Either $\ulcorner b \& k \urcorner \vdash h$ or $\ulcorner b \& h \urcorner \vdash k$ by Proof 3. Suppose $\ulcorner b \& k \urcorner \vdash h$. It is not the case that $\ulcorner b \& k \urcorner \vdash -h$ because it is not the case that $b \vdash \ulcorner -k \urcorner$ by assumption and (IR). Moreover, by (LT1), for any S, if $h \in S$, then S is not $MIS_{b \& k}$. Hence, $\mathrm{I}(h, k \& b)$. Suppose $\ulcorner b \& h \urcorner \vdash k$. Assume that it is not the case that $\ulcorner b \& k \urcorner \vdash \ulcorner -h \urcorner$ and that $j \in S$, $h \in S$, $j \neq h$ and S is $MIS_{b \& k}$. Then S is MIS_b by (LT4), and $(S \cap \overline{\{h\}}) \cup \{\ulcorner h \& k \urcorner\}$ is MIS_b by (LT3). If it is not the case that $\ulcorner b \& k \urcorner \vdash \ulcorner -h \urcorner$, then it is not the case that $b \vdash \ulcorner -h \urcorner$. Hence, if it is not the case that $\ulcorner b \& k \urcorner \vdash \ulcorner -h \urcorner$ and $j \in S$, $h \in S$, $j \neq h$ and S is $MIS_{b \& k}$, then it is not the case that $b \vdash \ulcorner -h \urcorner$ and there is an S' such that $j \in S'$, $\ulcorner h \& k \urcorner \in S'$, $j \neq h \& k$ and S' is MIS_b. From the initial assumption, $\mathrm{I}(h \& k, b)$ by Proof 3. Hence $\mathrm{P}(j, b) < \mathrm{P}(h \& k, b)$ by (IR). $\mathrm{P}(j \& k, b) \leqslant \mathrm{P}(j, b)$ by (PT2). Hence $\mathrm{P}(j \& k, b) < \mathrm{P}(h \& k, b)$. $\mathrm{P}(h, k \& b) = \mathrm{P}(h \& k, b) / \mathrm{P}(k, b)$ and $\mathrm{P}(j, k \& b) = \mathrm{P}(j \& k, b) / \mathrm{P}(k, b)$ by (P5). Therefore, $\mathrm{P}(h, k \& b) > \mathrm{P}(j, k \& b)$. Thus, if it is not the case that $\ulcorner b \& k \urcorner \vdash \ulcorner -h \urcorner$ and S is such that $h \in s$, $j \in s$, $h \neq j$, and S is $MIS_{b \& k}$, then $\mathrm{P}(j, k \& b) < \mathrm{P}(h, k \& b)$. Hence $\mathrm{I}(h, k \& b)$.

PROOF 5: If S is such that for any h, $\mathrm{I}(h, b)$ if and only if $h \in S$, then, if S is nonempty, it is not the case that $S \vdash \ulcorner -b \urcorner$. Assume S is such that for any h, $\mathrm{I}(h, b)$ if and only if $h \in S$ and S is nonempty. If S is nonempty, then it is not the case that $\vdash \ulcorner -b \urcorner$. By (IR), for any h, if $\mathrm{I}(h, b)$, then it is not the case that $h \vdash \ulcorner -b \urcorner$. Hence, if $S \vdash \ulcorner -b \urcorner$, then S must have at least two members. Suppose $S \vdash \ulcorner -b \urcorner$ and S has at least two members. There is a conjunction C of members of S such that $C \vdash \ulcorner -b \urcorner$. But, from Proof 3, $\mathrm{I}(C, b)$. If $\mathrm{I}(C, b)$, then $\mathrm{P}(C, b) > \mathrm{P}(-C, b)$. Hence $\mathrm{P}(C, b) > 1/2$ by (PT1). By (P3) and $b \vdash \ulcorner -C \urcorner$, $\mathrm{P}(C, b) = 0$. Thus $\mathrm{P}(C, b) = 0$ and $\mathrm{P}(C, b) > 1/2$. This is impossible. Therefore, it is not the case that $S \vdash \ulcorner -b \urcorner$.

PROOF 6: (IR) is equivalent to (IR_1). (IR_1) is as follows: $\mathrm{I}(h, b)$ if and only if (i) if it is not the case that $\ulcorner b\,\&\,h\urcorner \vdash k$, then $P(h, b) > P(k, b)$ and (ii) it is not the case that $b \vdash \ulcorner -h\urcorner$. Case one. If $I(h, b)$ by (IR), then $I(h, b)$ by (IR_1). Assume that (i) for any S, if $k \in S$, $h \in S$, $h \neq k$, and S in MIS_b, then $P(h, b) > P(k, b)$ and (ii) it is not the case that $b \vdash \ulcorner -h\urcorner$. Suppose it is not the case that $\ulcorner b\,\&\,h\urcorner \vdash k$. If neither $\ulcorner b\,\&\,k\urcorner \vdash h$ nor $b \vdash \ulcorner -k\urcorner$, then by (LT2), either $\{h, k\}$ is MIS_b or $\{h, k, \ulcorner -(h\,\&\,k)\urcorner\}$ is MIS_b, and hence $P(h, b) > P(k, b)$. On the other hand, if either $\ulcorner b\,\&\,k\urcorner \vdash h$ or $b \vdash -k$, then again, $P(h, b) > P(k, b)$. If $b \vdash \ulcorner -k\urcorner$, then since it is not the case that $b \vdash \ulcorner -h\urcorner$, $P(h, b) > P(k, b)$ by (P3). Suppose $\ulcorner b\,\&\,k\urcorner \vdash h$. $P(h, b) = P(h\,\&\,k, b)/P(k, h\,\&\,b)$ and $P(k, b) = P(h\,\&\,k, b)/P(h, k\,\&\,b)$ by (P5). But since it is not the case that $\ulcorner b\,\&\,h\urcorner \vdash k$ and is the case that $\ulcorner b\,\&\,k\urcorner \vdash h$, $P(k, h\,\&\,b) < P(h, k\,\&\,b)$ by (P3). Hence $P(h, b) > P(k, b)$. Hence, if it is not the case that $\ulcorner b\,\&\,h\urcorner \vdash k$, than $P(h, b) > P(k, b)$. This establishes case one. Case two. If $I(h, b)$ by (IR_1), then $I(h, b)$ by (IR). Assume that (i) if it is not the case that $\ulcorner b\,\&\,h\urcorner \vdash k$, then $P(h, b) > P(k, b)$ and (ii) it is not the case that $b \vdash \ulcorner -h\urcorner$. If $k \in S$, $h \in S$, $h \neq k$, and S is MIS_b, then, by (LT5), it is not the case that $\ulcorner b\,\&\,h\urcorner \vdash k$, and, hence, by assumption, $P(h, b) > P(k, b)$. Therefore, for any S, if $k \in S$, $h \in S$, $h \neq k$, and S is MIS_b, then $P(h, b) > P(k, b)$. This completes the proof of case two.

Proof 7: Consider a language L consisting of the members of a partition, $h_1, h_2, \ldots, h_n$, and truth functional combinations of such members. Let us call a statement which is a member of the partition or a disjunction of such members (each member occurring only once in a disjunction) a p-disjunction, and let D be a p-disjunction. By (IR), $I(D, b)$ if and only if, (i) for any p-disjunction D', if it is not the case that $\ulcorner b\,\&\,D\urcorner \vdash D'$, then $P(D, b) > P(D', b)$ and (ii) it is not the case that $b \vdash \ulcorner -D\urcorner$. Case one. If $I(D, b)$, then (i) for any p-disjunction D', if it is not the case that $\ulcorner b\,\&\,D\urcorner \vdash \ulcorner -D'\urcorner$, then $P(D, b) > P(D', b)$ and (ii) it is not the case that $b \vdash \ulcorner -D\urcorner$. Assume $I(D, b)$. By (IR), it is not the case that $b \vdash \ulcorner -D\urcorner$. Suppose D' is a p-disjunction and it is not the case that $\ulcorner b\,\&\,D\urcorner \vdash D'$. By Proof 6 it follows that $P(D, b) > P(D', b)$ which establishes case one. Case two. If (i) for any p-distinction D', if it is not the case that $\ulcorner b\,\&\,D\urcorner \vdash D'$, then $P(D, b) > P(D', b)$ and (ii) it is not the case that $b \vdash \ulcorner -D\urcorner$, then $I(D, b)$. Assume that (i) for any p-disjunction D', if it is not the case that $\ulcorner b\,\&\,D\urcorner \vdash D'$, then $P(D, b) > P(D', b)$ and (ii) it is not the case that $b \vdash \ulcorner -D\urcorner$. For any statement S of L, if it is not the case that $b \vdash \ulcorner -S\urcorner$, then there is some p-disjunction D' such that $\vdash \ulcorner S \equiv D'\urcorner$ by (LT6). If $\vdash \ulcorner S \equiv D'\urcorner$, then $P(S, b)$

$= \mathrm{P}(D', b)$ by (P2). Thus, (i) for any S, if it is not the case that $\ulcorner b \& D \urcorner \vdash S$, then $\mathrm{P}(D, b) > \mathrm{P}(S, b)$ and (ii) it is not the case that $b \vdash \ulcorner -D \urcorner$. Therefore, $\mathrm{I}(D, b)$ by Proof 6.

PROOF 8: Assume language L is as in Proof 7. If for any member of the partition $h_j (j > 1)$, $\mathrm{P}(h_j, b) > \mathrm{P}(h_1, b)$, then $\mathrm{I}(-h_1, b)$ by (IR). Assume that for any $h_j (j > 1)$, $\mathrm{P}(h_j, b) > \mathrm{P}(h_1, b)$. By (LT6), $\vdash \ulcorner -h_1 \equiv D_1 \urcorner$ where D_1 is the p-disjunction $(h_2 \vee h_3, \ldots, \vee h_n)$. For any p-disjunction D, if it is not the case that $\ulcorner b \& D_1 \urcorner \vdash D$, then $\mathrm{P}(D_1, b) > \mathrm{P}(D, b)$ by (P4) and the original assumption. Since by that assumption it is not the case that $b \vdash \ulcorner -D_1 \urcorner$, it follows from Proof 7 that $\mathrm{I}(D_1, b)$. Since $\vdash \ulcorner -h_1 \equiv D_1 \urcorner$, it follows from Proof 1 that $\mathrm{I}(-h_1, b)$.

PROOF 9: Assume language L is as in Proof 7. If $\mathrm{P}(h_n, b) > 1/2$, then $\mathrm{I}(h_n, b)$ by (IR). Assume $\mathrm{P}(h_n, b) > 1/2$. h_n is a p-disjunction and for any p-disjunction D', if it is not the case that $\ulcorner b \& h_n \urcorner \vdash D'$, then $\mathrm{P}(D', b) < 1/2$ by (P4) and (P1). Hence for any such D', $\mathrm{P}(h_n, b) > \mathrm{P}(D', e)$. Since it is not the case that $b \vdash \ulcorner -h_n \urcorner$ by (P3) and (P1), it follows from Proof 7 that $\mathrm{I}(h_n, b)$.

PROOF 10: Assume language L is as in Proof 7. If for any h_j and h_k, $\mathrm{P}(h_j, b) = \mathrm{P}(h_k, b)$, then $\mathrm{I}(S, b)$ only if $b \vdash S$. Assume that for any h_j and h_k, $\mathrm{P}(h_j, b) = \mathrm{P}(h_k, b)$, and that it is not the case that $b \vdash S$. If $b \vdash \ulcorner -S \urcorner$, then it is not the case that $\mathrm{I}(S, b)$ by (IR). Suppose it is not the case that $b \vdash \ulcorner -S \urcorner$. Then $\vdash \ulcorner S \equiv D \urcorner$, where D is some p-disjunction, by (LT6). By the same theorem it follows that there will also be some p-disjunction D' such that it is not the case that $\ulcorner b \& D \urcorner \vdash D'$ and is the case that the number of disjuncts in D is equal to the number in D'. Since the disjuncts are h_js, it follows that $\mathrm{P}(D, b) = \mathrm{P}(D', b)$ by (P4) and the assumption. Therefore, by Proof 7, it follows that it is not the case that $\mathrm{I}(D, b)$. Since $\vdash \ulcorner S \equiv D \urcorner$, it follows from Proof 1 that it is not the case that $\mathrm{I}(S, b)$.

University of Rochester

BIBLIOGRAPHY

[1] A. J. Ayer, 'Basic Propositions', in *Philosophical Essays*, London 1954, pp. 105–24.

[2] Rudolph Carnap, 'The Aim of Inductive Logic', in *Logic, Methodology and Philosophy of Science* (ed. by E. Nagel *et al.*), Stanford 1962, pp. 303–18.

[3] R. M. Chisholm, 'Theory of Knowledge', in Chisholm *et al.*, *Philosophy*, Englewood Cliffs 1964, pp. 261–86.

[4] R. Eberle, D. Kaplan, and R. Montague, 'Hempel and Oppenheim on Explanation', *Philosophy of Science* **28** (1961) 418–28.

[5] Gilbert Harman, 'Knowledge, Inference, and Explanation', *American Philosophical Quarterly* **5** (1968) 164–73.
[6] C. G. Hempel, 'Deductive-Nomological versus Statistical Explanation', in *Minnesota Studies in the Philosophy of Science*, Vol. III (ed. by H. Feigl and G. Maxwell), Minneapolis 1962, pp. 98–167.
[7] C. G. Hempel and P. Oppenheim, 'Studies in the Logic of Explanation', *Philosophy of Science* **15** (1948) 135–75.
[8] Risto Hilpinen, *Rules of Acceptance and Inductive Logic*, Amsterdam 1968.
[9] J. Hintikka and R. Hilpinen, 'Knowledge, Acceptance, and Inductive Logic', in *Aspects of Inductive Logic* (ed. by P. Suppes and J. Hintikka), Amsterdam 1966, pp. 1–20.
[10] David Hume, *An Inquiry Concerning Human Understanding* (The Library of Liberal Arts), New York 1957.
[11] Richard C. Jeffrey, 'Valuation and Acceptance of Scientific Hypotheses', *Philosophy of Science* **23** (1956) 237–46.
[12] Henry E. Kyburg, Jr., *Probability and the Logic of Rational Belief*, Middletown 1961.
[13] Henry E. Kyburg Jr., 'Probability, Rationality, and a Rule of Detachment', in *Proceedings of the 1964 Congress for Logic, Methodology and Philosophy of Science* (ed. by Y. Bar-Hillel), Amsterdam 1965, pp. 303–310.
[14] Keith Lehrer, 'Descriptive Completeness and Inductive Methods', *The Journal of Symbolic Logic* **28** (1963) 157–60.
[15] Keith Lehrer, 'Induction: A Consistent Gamble (Review of Levi, *Gambling With Truth*)', *Nous* **3** (1969) 255–297.
[16] Keith Lehrer and T. Paxson, 'Knowledge: Undefeated Justified True Belief', *Journal of Philosophy* **66** (1969) 225–237.
[17] Isaac Levi, *Gambling With Truth*, New York 1967.
[18] G. Maxwell, 'Scientific Methodology and the Causal Theory of Perception', in *Problems in the Philosophy of Science* (ed. by I. Lakatos and A. Musgrave), Amsterdam 1968, pp. 148–77.
[19] Karl Popper, *The Logic of Scientific Discovery*, London 1959.

REFERENCES

[1] W. V. O. Quine, W. F. Sellars, Gilbert Harman, and Paul Ziff are philosophers who have advocated some form of this theory.
[2] Cf. C. G. Hempel, P. Oppenheim [7] and Rolf Eberle *et al.* [4].
[3] This is due to David Kaplan.
[4] Cf. C. G. Hempel [6].
[5] This example is due to Rolf Eberle.
[6] Cf. A. J. Ayer [1].
[7] Cf. R. M. Chisholm [3].
[8] Cf. Grover Maxwell [18].
[9] For the author's views concerning such conditions see [16].
[10] David Hume [10], pp. 46–53.
[11] R. Carnap [2].
[12] *Ibid.*
[13] R. C. Jeffrey [11].

[14] Cf. I. Levi [17] and the author's review [15].
[15] See Karl Popper [19] and Risto Hilpinen [8].
[16] Gilbert Harman [5].
[17] C. G. Hempel [6], pp. 163–6, denies the conjunctivity of statistical explanation, but Hempel's argument is based on a principle of acceptance that leads to paradox as J. Hintikka and R. Hilpinen point out in [9].
[18] Such principles are discussed by Hempel, Levi, Popper, Hintikka, and Hilpinen in articles cited in the preceding footnotes. Both consistency and conjunctivity of hypotheses inductively inferred are rejected by H. E. Kyburg, Jr. in his article in this volume.
[19] The lottery paradox was first formulated, to my knowledge, by H. E. Kyburg, Jr., in [12], p. 197.
[20] H. E. Kyburg, Jr. rejects a principle of deductive closure in [13].
[21] Cf. I. Levi [17], especially Chapters 3 and 4.
[22] Cf. K. Lehrer [14].

ISAAC LEVI

PROBABILITY AND EVIDENCE*

I. SYNCHRONIC AND DIACHRONIC RATIONALITY

In 'The Aim of Inductive Logic', Rudolph Carnap has offered a clear statement of his conception of the way in which Inductive Logic or Confirmation Theory contributes to a theory of rational probability judgment.[1]

A theory of rational probability judgment is an integral part, according to Carnap, of a theory of rational decision making. Degrees of (subjective) probability, degrees of belief or degrees of credence, as Carnap calls them, are psychological dipositions to take risks of certain kinds and to avoid risks of other kinds. The degree of credence $\mathrm{Cr}_{X,T}(H)$, which an agent X assigns to a sentence H at a time T, is the highest betting quotient at which X is willing to bet on H at time T.

A psychological study of credence would examine the general laws relating degrees of credence assigned to different sentences (or propositions) by an agent X at a given time and the manner in which degrees of credence accorded by X to given sentences vary over time. The agent's credence function at a given time represents his 'state' of probability judgment at that time. It is a point in a credal configuration space. Over time, the agent moves through credal configuration space on a path determined by prior points and other factors (such as the acquisition of new evidence).

A psychological theory of credence can, therefore, be divided into two parts: (a) Synchronic theory concerned with general cross sectional laws determining relations between values of credence functions for different arguments at a given time, and (b) diachronic, dynamic or learning principles which relate credence functions at given times.

Carnap is not concerned with psychology but with a theory of rational credence as a system of norms or prescriptions controlling the manner in which men ought ideally to fix their probability judgments. He contends that such a theory makes prescriptions regarding how an agent ought to fix his credence judgments at a given time and regarding the manner in which he should revise them over time. In other words, Carnap's normative theory is intended to parallel psychological theory in having a synchronic and a diachronic part.

Thus, Carnap invites us to think of a theory of rational credence as providing a blueprint for the design of a robot who at every moment of time would conform to the requirements of synchronic rationality and would revise his credence judgments over time in accordance with the dictates of diachronic rationality. By thinking about such robots, advice can be found for human beings in "their effort to make their decisions as rational as their limited abilities permit".[2]

Carnap's conception of the aim of a theory of rational probability judgment is highly misleading. If taken seriously, it presupposes that an ideally rational man is one who *at every instant of time* (or, at least, during every minimal interval of time) conforms to the conditions of synchronic rationality and whose judgments change from instant to instant in accordance with the dictates of diachronic rationality.

In order to construct a theory of this sort which will be applicable in real life, heavy reliance will have to be placed on our knowledge of human capacities for learning. Without more knowledge of the psychological processes involved in fixing credence judgments than we actually have, we have no grounds for claiming that any theory of diachronic rationality which we propose will be one to which ordinary mortals can conform even to a good degree of approximation.

Patrick Suppes has clearly and effectively argued this point. Although he apparently agrees with Carnap's conception of the aim of a theory of rational probability judgment and the image of an ideally rational man underlying it, he regards Carnap's own theories of rational probability judgment as inadequate:

> The difficulty and subtlety of characterizing the mechanisms of information selection and at the same time a recognition of their importance in determining the actual behavior of men make me doubt that the rather simple Carnapian conception of inductive logic can be of much help in developing an adequate theory of rational behavior. Even the more powerful Bayesian approach provides a very thin and inadequate characterization of rationality because only one simple method for changing beliefs is admitted. It is my own view that there is little chance of defining an adequate concept of rationality until analytic tools are available to forge a sturdy link between the selection and use of evidence and processes of concept formation.[3]

This passage together with several other remarks clearly indicate that Suppes recognizes the psychologistic implications of Carnap's conception of rationality.[4] Any adequate diachronic theory of rational probability judgment must, in order to be applicable, be such that human beings can conform to its dictates. Such applicability can be determined only through studies in empirical psychology. At present, we do not have enough knowledge of such matters upon which to construct an adequate diachronic theory of rationality. Consequently, students of rational probability judgment and inductive logic in Carnap's sense ought to redirect their attention to learning theory. Suppes shares Carnap's image of an ideally rational man (or seems to) and bravely accepts the psychologistic consequences.

Those of us who are more timid see in Suppes' arguments a reductio ad absurdum. Carnap's image of an ideally rational man as one who meets conditions of rationality at every instant of time and revises his judgments in accordance with diachronic conditions from instant to instant is questionable precisely because it has the implications which Suppes sees so clearly.

But what would an alternative approach to Carnap's be like? Suppose that an agent indicates commitment to a given system of credence judgments. Either he or someone else notices that they deviate from certain conditions of rationality. The agent is not considered irrational for having deviated from these conditions. Perhaps, he was unaware of his deviation. Moreover, even if he were aware of it, he might not have been in a position to determine how he should adjust his beliefs in order to conform. But having recognized a deviation, he should, if he is rational, be committed to looking for a way to conform to standards of rationality.

Thus, recognition of a deviation from standards of rationality is not a sign of irrationality on the part of an agent unless the agent refuses to recognize any point in engaging in deliberations designed to rectify the deviation. A recognized deviation is a sufficient reason for embarking upon an inquiry. Consequently, conformity to conditions of rationality is a necessary condition for regarding one's situation as nonproblematic and inquiry as pointless.

Thus, principles of rationality can be regarded as stating necessary conditions for being in a certain kind of equilibrium in which critical review of one's probability judgments is unreasonable. These conditions are not sufficient; for one's beliefs might conform to conditions of rational probability judgment and yet one might inquire in order to obtain new information (or for other motives). Nonetheless, to consider principles of rational probability judgment as conditions of 'rational equilibrium' suggests a quite different conception of the aim of a theory of rational probability judgment than that implied by Carnap and Suppes.

The chief point is that the distinction between synchronic and diachronic rationality loses its significance. All conditions of rationality are analogous to equilibrium conditions. In a sense, they are synchronic conditions; but they are not conditions which an agent must satisfy at every instant of time. Furthermore, in stating conditions of rational equilibrium, no prescription is made regarding the psychological path to be taken in moving from disequilibrium or from one equilibrium position to another. In other words, there are no norms prescribing rational learning processes.

Such a theory of rational probability equilibrium has, nonetheless, something to say about how probability judgments are to be revised in the light of new evidence. Consider the 'neoclassical' or 'marginalist' theory of consumer demand. Given a consumer's utility function, his income and the prices which he faces, that theory provides an account of how the consumer would allocate his income to the purchase of various commodities if he is 'in equilibrium'. Moreover, if some of the independent factors (for example, the consumer's utility function and his income) remain constant but other factors (e.g., the prices) change, the theory specifies what the new equilibrium demand would be. The theory can not predict, however, that if the prices do change under the circumstances described the consumer will attain the new equilibrium position or if he does how long the process of readjustment will take and what the behavioral and psychological properties of the readjustment process will be like. The theory of how demand responds to price changes is, in widely used jargon, a 'comparative static' theory which studies shifts from one equilibrium position to another.[5]

According to the view which shall be adopted here, a theory of rational probability revision is a normative analogue of such a comparative statical theory. Obtaining new evidence is analogous to a change in prices. The theory determines a new position of rational equilibrium when additional evidence is obtained. It makes neither predictions nor recommendations as to how the equilibrium is to be attained but only recommends that ceteris paribus it should be.

To enjoin an agent to move toward rational equilibrium presupposes that the agent is able to do so. 'Ought' does after all imply 'can'. Thus, psychological considerations are, as Suppes insists, relevant to appraising the applicability of a theory of rational probability judgment even when interpreted as a comparative statical theory. But the abilities ascribed to an agent are of a rather different order on a comparative statical view than those which a theory prescribing the details of the psychological processes involved in attaining equilibrium presuppose. Indeed, judgments as to the feasibility of attaining equilibrium by some path or other can quite frequently be made on the basis of rather casual observation of human behavior.

For example, it does not take great learning in psychology to recognize that if a Bayesian theory of rational probability judgment is adopted then in some situations an agent will be incapable of making the calculations required to determine what his equilibrium is to be.

The same type of situation also suggests that the remedies for such difficulties in the application of Bayesian theories are not likely to be found by concentrating primary effort in the study of learning theory. Improved computer technology which will cheapen the cost of calculations looks more promising. In any event, if learning theory will be of help here, it is not likely that it will do so by pointing the way to a better theory of rational probability judgment but by devising techniques for enabling agents to calculate more quickly and exploit their memories more effectively. One ought not to be dogmatic here; but, at the present, Suppes' exhortation to students of probability and induction to bend their efforts to a study of learning remains unconvincing.

These remarks about the relations between studies of rational probability judgment and psychology are predicated on the assumption that students of rational probability judgment should discard Carnap's conception of such a theory as providing a design for a perfectly rational robot or learner characterizing in some detail the processes involved in changing probability judgments in response to the acquisition of new evidence or sensory stimulation.

Among the casualties resulting from the abandonment of this approach are principles of conditionalization construed in a diachronic, kinematic or learning theoretic manner. So is the surrogate for conditionalization proposed by Richard Jeffrey. However, alternative formulations of conditionalization principles might be found which are compatible with an equilibrium and comparative statical approach to rational probability judgment. Such principles would be relatively immune to the objections raised by Suppes against diachronic principles of conditionalization. In subsequent sections, attention will be given to several equilibrium and comparative statical versions of conditionalization.

II. SUBJECTIVISM

Following Carnap, the judgments made by an agent X at a given time T regarding fair betting quotients will be represented by what he calls a 'credence function' $\mathrm{Cr}_{X,T}(H)$. We shall assume, for the present, that the credence function is defined for all sentences in X's 'language' L – whatever that is.

The least controversial requirement for rational probability or 'credence' judgment is the following:

(I) $\mathrm{Cr}_{X,T}(H)$ is strictly coherent.

As is well known, the requirement of strict coherence implies that the credence function at a given time must be a regular probability measure.[6]

According to the comparative statical approach adopted here (I) is a necessary condition for rational equilibrium. If a violation of coherence requirements is detected in X's credence judgments, he need not be considered irrational provided that he recognizes the need to readjust his judgments to avoid the violation.

Let K be the set of sentences which X accepts as evidence at T. According to (I), if X is in equilibrium, each element of K is accorded a credence value of 1. Moreover, K is a deductively closed set. We do not need to suppose that each member of K is explicitly listed. There are other ways in which X can characterize K – for example, by listing some subset of K and stating that K consists of the deductive consequences of items on the list.

One important feature of condition (I) is that if H is any sentence neither in K, nor inconsistent with an element in K, then for every real number r between 0 and 1 there is a credence function satisfying (I) which assigns

r to H. If (I) were the sole condition for rational equilibrium, the evidence available to an agent X would impose only very weak constraints on equilibrium judgments of credence for sentences whose truth or falsity is not deducible from the evidence. Consequently, one could not exploit empirical information to revise credence judgments.

Suppose that at T, X's total evidence is K. Between T and T', he 'acquires' in some sense additional evidence E so that his new total evidence becomes $K\&E$. What should his credence function, in equilibrium, be like at T'?

> Let X's function $\mathrm{Cr}_{X,T}(H)=f(H)$ having the following properties:
> $f(H)=r$
> $f(H/E)=f(H\ \&\ E)/f(E)=k.$

According to a widely held view, $\mathrm{Cr}_{X,T'}(H)$ should equal $f(H/E)$ $=\mathrm{Cr}_{X,T}(H/E)$. The principle upon which this is usually based is the diachronically interpreted principle of conditionalization. However, there is a way to construct a principle of conditionalization similar to this one acceptable within the framework of a comparative statical theory.

(II) If $\mathrm{Cr}_{X,T}$ is an equilibrium function at T when K is the total evidence and E alone is acquired between T and T', then $\mathrm{Cr}_{X,T'}$ is a new equilibrium credence function if and only if for every H, $\mathrm{Cr}_{X,T'}(H/E)=\mathrm{Cr}_{X,T}(H/E)$.

Given that (I) implies that $\mathrm{Cr}_{X,T'}(E)$ must in equilibrium be equal to 1, (II) implies that $\mathrm{Cr}_{X,T'}(H)=\mathrm{Cr}_{X,T}(H/E)$.

Observe that (II) does not recommend that $\mathrm{Cr}_{X,T'}(H)$ equal $\mathrm{Cr}_{X,T}(H/E)$, but only that the equality obtain if X's credence judgments at T' are in rational equilibrium.

Now if the total evidence K at T does not uniquely determine the equilibrium credence function adopted by X at T, condition (II) does not by itself determine uniquely how credence judgments ought to change.

Suppose, for example, that not only is $f(H)$ a possible equilibrium credence function at T but that $g(H)$ is also an equilibrium function, where $g(H)=r'$ and $g(H/E)=k'$. Assume that X actually was in equilibrium at T and adopted $f(H)$ as his credence function. Condition (II) does not prevent X from adopting $g(H/E)$ as his equilibrium credence function at T'.

The same point can be made from a different direction. If X did not obtain any more evidence after T but remained with the same total evidence K, this does not mean that X cannot move from one equilibrium position to another. He can, according to (I) and (II), adopt $f(H)$ as his credence function at one time and $g(H)$ at another.

This result cannot be avoided by strengthening (II) with an additional rider which constrains X to stick with the equilibrium credence function which he first adopts unless his evidence changes and to conditionalize relative to that credence function when he obtains new evidence. Why should X be saddled with his past commitments in this way when, at the time when he made the commitment, it would have been perfectly rational for him to have adopted other commitments?

Writers on probability of subjectivist and personalist persuasion have often contended that they are able to accommodate the idea that empirical evidence does contribute to determining credence judgments. Yet, they insist that two agents sharing the same evidence can both be in rational equilibrium and have different – indeed, radically different – credence functions. They attempt to escape from extreme subjectivism here precisely by adopting a principle of conditionalization which is at worst untenably diachronic and at best saddles agents with their past judgments in a manner which cannot possibly be defended on their own principles.[7]

In my opinion, if we are to take seriously the idea that probability – i.e., credence – judgments ought to be revisable in the light of empirical evidence, we must be prepared to admit that empirical evidence does in some sense objectively confirm or disconfirm hypotheses. Thus, in spite of the many important contributions which personalists have made to probability theory, we must regard their position in the final analysis as excessively subjectivist and utterly incompatible with an empiricist epistemology.

One further point ought to be made before concluding this section. Condition (II) by itself adds no additional strength to condition (I). If the truth or falsity of H is not entailed by either K or $K\&E$, then for every r between 0 and 1 there is an equilibrium credence function when the evidence is K and an equilibrium function when the evidence is $K\&E$ such that H is accorded the credence value r by both credence functions. This emphasizes in another way the critical point already made. Unless we can add equilibrium conditions which specify the way in which the total evidence K constrains the credence function, no comparative statical theory can offer an account of the way additional evidence alters the equilibrium credence function.

III. ON TOTAL EVIDENCE

One way in which one might hope to develop a suitable theory is through a variant of Carnap's own approach. If a measure of confirmation $c(H, K)$ suitable for X's language can be constructed, then the following equilibrium condition can be introduced.

(III) If X's total evidence at T is K, then X's equilibrium credence function $\mathrm{Cr}_{X,T}$ at T is such that $\mathrm{Cr}_{X\ T}(H) = c(H, K)$.

Observe that (III) is, in a sense, a form of conditionalization. Yet, it is neither diachronic nor does it involve explicit comparisons of credence functions in equilibrium at different times. Yet, it surely permits a rather powerful comparative statical theory. For example, if at T, X's total evidence is K and at T' is $K\&E$, it does imply that $\mathrm{Cr}_{X,T'}(H)$ should equal, in equilibrium, $\mathrm{Cr}_{X,T}(H/E)$. Here the objections raised previously do not hold; for the equilibrium positions both at T and T' are uniquely determined by the total evidence so that this relation obtains.

But this Carnapian approach has some well-known difficulties of its own. The notion of an a priori determined confirmation function is unacceptable on rather solid grounds to many who, though committed to the view that confirmation relations are objective, consider them to be empirical as well. Furthermore, Carnap has faced considerable technical difficulties in constructing suitable confirmation measures for even moderately rich languages.

But even if these objections and difficulties could be overcome, Carnap's confirmation functions would have no applicability in real life; for no one could ever under any realistic circumstances attain rational equilibrium. In order to attain rational equilibrium, X would have to determine what his equilibrium credence function is. To do this, he would have to identify his total evidence in such a manner as to allow him to use Carnap's confirmation function via (III) to determine his credence function. No one can review his total evidence in the manner required.[8]

In what sense does Carnap's theory require that the total evidence be explicitly given before his confirmation measures are applied? The total evidence must be characterized as the deductive consequences of a specific list of axioms or of a set of axioms characterized by axiom schemata. It is only for evidence cast in some form such as this that Carnap's measures enable one to determine degrees of confirmation for hypotheses. And it is precisely such explicit statement of total evidence that no ordinary agent is able to provide.

There are, however, other ways of characterizing the total evidence which permit (III) to apply. X's total evidence at T might be described as confirming H to degree r. That is to say, the metalinguistic predicate $c(H, -) = r$ is declared to be true of the deductively closed set of sentences constituting X's total evidence. There may be several deductively closed sets of sentences in the language under consideration of which this predicate is true. However, in order to apply condition (III), it is not necessary to identify which of these closed sets constitutes X's total evidence. Since X's total evidence is trivially X's total evidence and $c(H, X\text{'s total evidence}) = r$, condition (III) recommends that in equilibrium $\text{Cr}_{X,T}(H) = r$.[9]

Suppose that X cannot declare that $c(H, X\text{'s evidence at } T) = r$, but only that $c(H, X\text{'s evidence at } T)$ equals either r or s or falls in the interval between r and s, or is greater than $c(G, X\text{'s evidence at } T)$. In such an event, condition (III) does not prescribe a unique credence value for H at T. Nonetheless, it does constrain the credence value – e.g., in the first case, to the values r and s. There is an element of arbitrariness in the choice of a credence function here as long as the only characterization of X's total evidence is of the type given. But this concession to subjectivism is not excessive. After all, it may be possible for X to obtain more evidence and thereby constrain his credence judgment more definitely. If he has the opportunity of looking for such evidence before using his credence function to make a decision, he should do so. Otherwise, he will, indeed, have to select a credence function arbitrarily. But a theory of rational equilibrium should recognize that if an agent is not able to attain a suitable equilibrium, he may have to adopt a 'second best' approach and that such an approach might involve some arbitrary choice. A theory which insists that the scope of arbitrary choice, can, in principle, be indefinitely reduced recognizes this but does not, therefore, revert to extreme subjectivism.

But isn't the escape from subjectivism bogus? Suppose that X declares his background to be a K such that $c(H, K)=r$, whereas Y declares his background to be K' such that $c(H, K')=s$. Is there any objective way to settle their differences?

Even if it were true that what is counted as evidence is an arbitrary affair, the subjectivism which results is different from the subjectivism popular among Bayesians according to which given evidence K and a sentence H not entailed by K for every r between 0 and 1 there is a legitimate credence function which accords H the value r. According to the idea suggested here, credence judgments would be constrained by empirical assumptions.

Moreover, there is surely no need to grant without further ado that what is admitted as evidence is an arbitrary matter.

Nonetheless, the objection carries some weight – or seems to. To claim that the evidence is K such that $c(H, K)=r$ is, indeed, an empirical claim. But unless the set K is recharacterized in a manner which allows its contents to be explicitly given, it is not clear that the truth or falsity of members of K can be checked by confrontation with observation reports. Given that no one can render his total evidence K explicit in this manner, assumptions about empirical evidence of the sort considered here are immune to serious empirical criticism.

Consider, however, that even though X cannot explicitly characterize his total evidence, he can surely explicitly cite some of it. Moreover, there is no relevant a priori or psychological constraint on which part he can render explicit and which he cannot. Our inability to explicitly review all of our empirical evidence all at once means, therefore, that we cannot bring all of the empirical assumptions upon which we base our credence judgments under critical scrutiny at the same time. But it does not mean that the total evidence is immune from criticism in a piecemeal fashion.

Thus, on many occasions it would be inappropriate for X to characterize his total evidence as a set of sentences K such that $c(H, K)=r$. If his credence judgments are being challenged, he might be required to cite explicitly evidence upon which his judgments are based. Let his explicitly cited evidence consist of propositions expressed by sentences E. X would then characterize his total evidence as consisting of the deductive consequences of the conjunction of E and B such that $c(H, B\&E)=r$. He might further characterize his implicit or background information B by identifying values of $c(H, B)$, $c(E, B)$, etc.

X's argument, therefore, in defense of a given credence judgment $\mathrm{Cr}_{X,T}(H)$ can then be reconstructed as follows:

(1) $c(H$, the deductive consequences of E and X's background information$) = r$.

(2) X's total evidence at T consists of the deductive consequences of E and X's background evidence at T.

(3) $\mathrm{Cr}_{X,T}(H) = r$.

This mode of argument renders pointless any concern about constructing confirmation functions. Given that we cannot explicitly characterize total evidence, we cannot directly apply them; and in order to make evidential assumptions sufficient to justify credence judgments, we do not need them.

Certain assumptions must, of course, be made about confirmation functions. Conditions (I) and (III) imply that confirmation measures must be probabilistic in structure. Moreover, it is assumed that statements such as $c(H, K) = r$ for explicit K and H are objectively true or false. This implies no more than that evidence objectively confirms hypotheses to various degrees.

Notice that we are free to regard statements like '$c(H, K) = r$' for explicit H and K as empirical. To be sure, if we follow the procedure adopted here and regard confirmation functions as taking sentences and sets of sentences in a given language L as arguments, then in the typical case confirmation statements will not be expressible in L. The empirical assumptions they assert are not part of the total evidence expressible in L. But L can be enlarged to include such confirmation statements when the demand arises. Or, alternatively, L can be enlarged to include statements K' such that for the confirmation function c' defined for the enlarged L', $c'(H, K \& K') = r$. In effect, Carnap's confirmation functions can always be viewed as conditional on evidence not explicitly expressible in the languages for which these functions are defined.

Thus, even though no one can explicitly cite his total evidence, there does not seem to be an a priori or a psychological bar preventing an investigator from explicitly citing part of it and from delving further into the implicit background information B, thereby enlarging the explicit part as the occasion requires. We do not need to conclude that any of X's empirical or confirmational assumptions are immune to explicit empirical criticism.

These remarks have a direct bearing on how credence judgments are revised in the light of new evidence. Suppose that at T X accords H a credence value of r and $\mathrm{Cr}_{X,T}(H/E)=s$, where E is not part of X's evidence.

If X is taken to be in rational equilibrium at T, his total evidence X must be such that $c(H, K)=r$ and $c(H, K\&E)=s$. By acquiring new evidence E subsequent to T, X is like the consumer who faces a change in prices. He has been thrown out of equilibrium and must find a new equilibrium position.

If at T' subsequent to obtaining E, *X does not retract any of the evidence he accepted at* T, he must regard his total evidence to be $K\&E$, where K is such that $c(H, K\&E)=s$. By condition (III), in equilibrium at T', $\mathrm{Cr}_{X,T'}(H)$ should be equal to the equilibrium $\mathrm{Cr}_{X,T}(H/E)$. This is the result recommended by customary principles of conditionalization. Observe, however, that this recommendation is not based upon a general principle of rationality prescribing that new credence judgments should equal past conditional credence judgments. A crucial rider has been added. It requires that such a result obtain only if past evidential assumptions have not been modified. Since, on some occasions, such modifications are legitimately called for, conditionalization cannot be imposed as a general principle for revising credence judgments in the light of new evidence.

Two further points deserve repetition here. Equilibrium condition (III) is a version of conditionalization relating confirmation measures to credence measures. It is an equilibrium condition, however, lacking any temporal features.

The second point is that even when X does not retract any of the evidence he accepted at T, $\mathrm{Cr}_{X,T'}(H)$ need not always equal in equilibrium $\mathrm{Cr}_{X,T}(H/E)$. For X may judge (both at T and T') that K is such that $c(H, K\&E)$ equals either r or s or that it falls in an interval between r and s. In that event, X is free to adopt $\mathrm{Cr}_{X,T}(H/E)=r$ and $\mathrm{Cr}_{X,T'}(H)=s$.

Thus, situations where the equilibrium positions at T and T' warrant equating $\mathrm{Cr}_{X,T'}(H)$ with $\mathrm{Cr}_{X,T}(H/E)$ are very special. They arise when the total evidence at T determines a unique equilibrium credence value for H given E and the addition of E to the total evidence between T and T' does not involve abandoning any of the evidence accepted at T.

IV. EMPIRICISM AND FALLIBILISM

According to the approach adopted here, the manner in which credence judgments are to be revised depends critically on the empirical assumptions which the agent makes. The totality of such assumptions cannot be explicitly cited by the agent all at once or in any reasonable period of time. However, any portion of the total evidence which is not rendered explicit can in principle be made so. Consequently, an agent's credence judgments are empirically arguable – i.e., they can be critically evaluated through an examination of some of the empirical assumptions upon which they are based.

Of course, to adopt such a view presupposes that sometimes it is legitimate for an agent to accept empirical propositions as evidence and accord them a credence value of 1. If no empirical propositions could legitimately be accepted as evidence, then all credence values would perforce be based upon a priori assumptions alone. If a priori evidence suffices to warrant assigning definite credence values to empirical hypotheses, then for all time there is only one equilibrium position. If, on the other hand, a priori evidence permits arbitrary choice between several equally legitimate credence functions, that choice remains the same for all time. Both of these views are untenable. It must be legitimate, therefore, to accept empirical statements as evidence.

This implication is one which many writers find difficult to concede. To accept E as evidence is to accord it a credence value of 1. This means that the agent X regards acting on E as without risk. Many philosophers consider this reasonable only if X has an infallible warrant of the truth of E. If it is agreed that no empirical statement can be provided with such a warrant, no empirical statement should rationally be accepted as evidence.[10]

This objection is predicated upon a dubious demand for infallibility. We may, I think, quite legitimately regard acting on E as being without serious risk even though we recognize the theoretical possibility that E is false. The fact that E might be false is not, in itself, sufficient to deprive us of the right to regard E as certain. It does, however, oblige us to keep an open mind regarding the truth of E – i.e., to be prepared to remove E from its evidential status when a serious challenge is made to according it that status. The difference between a priori evidence and empirical evidence is not to be found in differences in credence values assigned to the a priori and the empirical but in the corrigibility of empirical evidential assumptions. The a priori consists of that which is immune to revision in the face of 'recalcitrant experience'. The empirical is more sensitive. This difference between the empirical and the a priori in no wise implies that in a given equilibrium state empirical and a priori assumptions are distinguishable with respect to the credence values assigned them. When engaged in deliberation for the purpose of making a choice between alternative policies, we regard acting on *E* as without risk if there is no serious doubt as to the truth of *E*. 'Cartesian' doubts regarding the truth of *E* do not count.

These remarks are nothing more than a brief restatement of the fallibilist position advocated by Pierce and Dewey. Fallibilism is not to be confused with the banal view that no empirical statement can be provided with an infallible warrant. Rather it is the position which claims that we are sometimes entitled to accept empirical statements as evidence and, hence, as true even though we lack such an infallible warrant. I have attempted to argue that fallibilism is the only tenable view on the basis of which a theory of rational probability judgment can be constructed which regards credence judgments as being grounded empirically. However, Richard Jeffrey, who is an articulate spokesman for the antifallibilist view, has recently proposed an account of rational probability judgment which claims to require that credence judgments be empirically grounded without necessitiating the admission of any empirical statements as evidence.[11]

Jeffrey's approach involves making a distinction between basing credence judgments on empirical evidence and basing credence judgments on direct responses to sensory stimulation.

When *X* makes an observation and admits an observation report *E* into evidence, we can regard his accepting *E* as evidence and according it a credence value of 1 as a causal response to the sensory stimulation. Jeffrey contends that one can respond to sensory stimulation by shifting a credence assignment to *E* from one nonextreme value to another. Indeed, he argues that even in those cases where *X* apparently accepts *E* as evidence in direct response to sensory stimulation, we can regard him as assigning *E* a credence value so close to 1 'as to make no odds'. Strictly speaking, a rational *X* never assigns a credence value of 1 to any empirical statement. Yet *X*'s credence judgments are empirically based in the sense that some of them are direct responses to sensory stimulation and the others are determined through conditions of rationality by those which are direct responses.

According to Jeffrey, the salient feature of credence judgments made in direct response to sensory stimulation is that the agent is compelled by the stimulation to make those judgments. Sensory stimulation in no wise justifies the responses but causes them to occur. Such compulsory credence judgments provide the empirical basis for all probabilistic belief.

Jeffrey does not maintain that compulsory credence judgments of the sort mentioned are invariably sound or legitimate. Nor need they be permanent. "New experience or new reflection (perhaps, prompted by the arguments of others) may lessen the bonds of compulsion, and may then establish new bonds; and the effect may be that the new state of belief is sounder than the old, or less sound."[12]

Why are compulsory responses reliable?

> Quite simply put, the point is: coming to have suitable degrees of belief in response to experience is a matter of training – a *skill* which we begin acquiring in early childhood, and are never quite done polishing. The skill consists not only in coming to have appropriate degrees of belief in appropriate propositions under paradigmatically good conditions of observation, but also in coming to have appropriate degrees of belief between zero and 1 when conditions are less than ideal.[13]

Consider two agents *X* and *Y* who have observed a piece of cloth. *X* observes the piece of cloth under good light and responds by according the statement that the cloth is green a credence value of .999999. *Y* observes the cloth under poor light but responds by assigning *G* the same credence value. Both respond – indeed are compelled to respond – in the same way to different stimuli. Yet, the judgment of one of them, according to Jeffrey, would not be as sound or legitimate as that of the other.

If neither *X* nor *Y* accept any empirical assumptions as evidence, neither of them can render a verdict as to the reliability of their credence judgments. Thus, neither *X*, *Y*, nor anyone else who refuses to accept empirical evidence could criticize *Y*'s response by noting that it was made under poor light. Indeed, *X* could not defend his response on the grounds that it was a response to sensory stimulation made under satisfactory conditions. To do so requires admission of empirical assumptions as evidence.

Of course, if no one was concerned with critical evaluation of credence judgments, this would not matter. If human beings had perfected their skills in responding to sensory stimulation so that they were invariably reliable, perhaps, there would be no need for criticism. As in the case of the angels, being rational would be doing what comes naturally. But to prescribe for a community of angels is like designing one of Carnap's perfectly rational robots. It is to consider a rational agent as one conforming to conditions of rationality at every instant of time. Clearly, this approach is unacceptable. We are not angels nor are we such robots. Our skills are imperfect and our responses often less than reliable. Criticism is important.

But our credal responses to sensory stimulation cannot be criticized if we make no claims about them. If we do not claim that they are responses to sensory stimulation, we are not even committed to saying that they are empirically grounded in the sense that Jeffrey requires them to be. They could equally as well be compulsory responses to changes in metabolism. Moreover, if we recognize a distinction between reliable and unreliable responses to sensory stimulation, as Jeffrey himself does, we must defend our credal responses by claiming that they are made under satisfactory conditions. Otherwise, no commitment is made regarding the satisfactoriness of the judgment at all. The response remains immune to relevant criticism.

Jeffrey seems to think that the fact that credal responses to sensory stimulation are compulsory is important. Being compulsory, we cannot help having these responses. We must, if we are rational, adjust our other judgments to fit in with these compulsory responses.

> In my scientific and practical undertakings I must make use of such compulsory beliefs. In attempting to understand or to affect the world, I cannot escape the fact that I am a part of it. I must rather make use of that fact as best I can.[14]

Of course, we must do what we must do. This is true of credal responses to changes in blood pressure or of unreliable responses to sensory stimulation as much as it is of reliable responses. There would be little point in criticism if once one is compelled to credence judgments, one was compelled to stick with the judgments indefinitely. But if such judgments can appropriately be criticized, it must in general be true that such responses can be discarded in due time. If so, we are not constrained to make use of all of our compulsory beliefs. When we are faced with making decisions at the very instant when the beliefs are compulsory, we cannot help ourselves. In general, however, this will not be the case – at least for those cases where criteria of rationality have any applicability. Such criteria are designed for situations where an agent is capable of deliberation and can, in due time, remove the shackles of compulsion.

To be sure, an empiricist will require that some items admitted into evidence or some credence judgments should be based on reliable responses to sensory stimulation. But it is not the compulsory character of such responses which is decisive. It is of far greater importance that the responses be to sensory stimulation and, still more important, that they be reliable responses.

Perhaps, Jeffrey does not mean to deny that agents do admit empirical assumptions into evidence. He may only intend to deny that evidence is ever explicitly stated by citing sentences translatable into some suitably formalized language *L*. Thus, when *X* assigns a credence value of .999999 to *G*, he may declare that he has made a response under satisfactory circumstances to sensory stimulation. He has admitted a certain assumption into evidence; but he has not and may not be capable of rephrasing that assumption in a manner which indicates in more specific form what the conditions are. Jeffrey's point would then be that an agent can no more completely characterize the conditions of observation than he can explicitly cite his total evidence.

If this is, indeed, Jeffrey's point, it has the same merits as the observation that no one can explicitly cite his total evidence. This is indeed true. However, it is also true that one can explicitly describe some features of the conditions of observation. Moreover, the precision with which they are described is not limited in any a priori way; and if responses to sensory stimulation are to be critically evaluated, some explicit description of the conditions of observation will be required.

Thus, it seems to me that not only must an agent tacitly accept empirical statements as evidence. He must be capable of rendering some part of what is implicit explicit.

In sum, it seems to me that Jeffrey has failed to drive the wedge between responses to sensory stimulation and admitting empirical statements into evidence which would be required to avoid allowing agents to accept empirical statements into evidence and still retain the claim to an empiricist epistemology.

Empiricists must admit that some empirical statements can legitimately be accepted as evidence and accorded a credence value of 1. If they also admit that no empirical statements can be afforded an infallible guarantee, they must also adopt a fallibilistic point of view. Jeffrey's shrewd attempt to avoid a commitment to fallibilism does not succeed in nullifying this verdict.

V. ON RATIONAL ACCEPTANCE

The fallibilist and empiricist position adopted here insists that it is sometimes legitimate to accept as evidence and, therefore, to assign a credence value of 1 to empirical statements for which no infallible guarantee can be given. Such statements are always open to theoretical doubt; for it is logically possible that they are false. But judgments of risk upon which actions are based do not take into consideration such theoretical doubts. Only serious doubts are relevant to determining risk. A proposition whose truth is not judged problematic at a given time is a proposition to which a credence value of 1 can legitimately be assigned. Acting on it is without serious risk.

Thus, when X accepts E as evidence he regards accepting E as true as nonproblematic. Such acceptance may be caused by sensory stimulation. But cultural and psychological factors also determine what is accepted as evidence. There is nothing a priori wrong with accepting statements as evidence in response to a wide variety of factors. As long as no serious challenge is raised against such acceptance, they remain nonproblematic.

However, no evidential assumption is immune to serious challenge. When such an assumption is questioned, a rational agent will remove it from his evidential corpus and attempt to determine on the basis of the remaining nonproblematic evidence whether the problematic item should be reinstated as nonproblematic evidence or replaced by some other item.

If we are to avoid subjectivism and relativism, we must assume that objective standards regulate revision of the evidential corpus. Objectivity does not require that our initial evidential corpus be determined by objective standards but that the standards whereby that corpus is revised and corrected be objective.

One way in which the evidential corpus is revised is through observation. When observations are made under conditions which are nonproblematically reliable, the statements accepted as evidence in response to sensory stimulation are nonproblematically accepted as evidence. New evidence obtained by observation can subsequently be questioned. It may be granted that observations made under conditions *C* are reliable but it may be seriously doubted whether the observations to which the report in question was a response were made under those conditions. Or the reliability of observations made under conditions *C* can be questioned. In both cases, whether the observation report is to be returned to the evidence or not can be settled only by inference from other evidence.

Much evidence does not, of course, consist of observation reports. Empirical generalizations and theoretical assumptions also appear in the evidential corpus – for example, in stating conditions of reliability of observation reports. Objective correction of such evidence again requires inference from evidence.

Thus, an important role must be assigned to criteria for rational acceptance which control conditions under which empirical statements can be added to the evidential corpus. Moreover, the criteria must include conditions for accepting statements *H* as evidence on the basis of evidence which does not entail *H*. To deny the relevance of criteria for inductive acceptance is to deny that evidential assumptions can be corrected and revised in an objective manner. Since, Richard Jeffrey to the contrary notwithstanding, rational credence judgments are based on empirical assumptions, to deny the possibility of objective criticism and revision of the evidential corpus is tantamount to denying that the empirical assumptions upon which probability judgments are based can be objectively criticised.

When can an empirical statement *E* be objectively and legitimately accepted as evidence via nondeductive inference from other evidence *K*? The short answer is when the other evidence *K* warrants regarding accepting *E* as true as nonproblematic. And accepting *E* as true is nonproblematic when *E* can legitimately be accepted as a correct answer to every problem or question which is recognized to be serious at the time when the status of *E* is under consideration.

From this point of view, we may distinguish three different senses in which a statement can be accepted as true via nondeductive inference from evidence *K*.

(i) *H* may be accepted as a true answer to a specific problem or question *P* for which *H* is a potential answer.

(ii) Sometimes *H* might be accepted as a true answer to a specific problem *P* but only tentatively so. This means that although *H* is the best answer on the evidence available, the evidence *K* might not so decisively settle the question in favor of *H* as to render further inquiry and the acquisition of additional evidence pointless. When, however, *H* is accepted as an answer to a specific problem *P* in a manner which renders pointless further inquiry to check on the propriety of adopting *H* as the correct answer to *P*, *H* is accepted as a final or terminal answer to *P*.

(iii) Both acceptance as a true answer to a specific problem *P* and acceptance as a final true answer to a specific problem *P* are relativized not only to the total evidence but to the problem *P*. In order to accept *H* as evidence, *H* must be accepted as a final answer relative to all problems which are regarded as serious problems at the time when such acceptance is being considered and for which *H* is a potential answer. If *H* is accepted as final relative to *P* but is not so accepted relative to *Q* for which *H* is a relevant answer, *H* cannot be accepted as evidence. In that case, the truth of *H* remains subject to serious doubt and is questionable.

This conception of the conditions under which an empirical hypothesis might legitimately be accepted as evidence via inference from other evidence is not intended as a final characterization. Rather it is to be regarded as the framework for guiding research into certain important questions pertaining to inductive inference.

In particular, its adoption suggests that a central objective of an account of scientific inquiry is to provide an analysis of the invariant features of specific inquiries where the aim is to provide true answers to given questions. This will include devising criteria for determining what are potential or relevant answers to specific problems (abduction), planning the acquisition of new evidence, criteria for weighing evidence in order to decide which of the potential answers ought to be accepted as true relative to the available evidence, criteria for deciding whether the answer accepted is final or whether additional inquiry is in order. Finally, it suggests that some attention be given to considering differences between serious and nonserious problems and questions.

Thus, criteria for rational inductive inference are to be studied within the framework of an account of rational problem solving. Indeed, an inductive inference is an attempt to justify a solution to a problem of a specific kind – namely, to a cognitive problem where the aim is to find a correct answer to a given question. But an adequate account of rational problem solving relevant to cognitive problems should be a specialization of an account of rational problem solving applicable to practical decision problems as well. And Bayesian accounts of rational probability judgment are intended to be an integral part of a general account of problem solving. Thus, we have come full circle. I have argued that a theory of rational probability judgment which allows that probability judgments are empirically grounded must make room for legitimately accepting statements as evidence. But, according to the view adopted here, an account of rational acceptance ought itself to be embedded in an account of rational problem solving which, in turn, will make use of an account of rational probability judgment.

Thus, those who would build a theory of rational probability judgment without taking into consideration questions about rational acceptance and those who would worry about acceptance while ignoring probability are all committed to one sided images of the rational conduct of inquiry. The fundamental philosophical problem pertaining to probability and induction is not whether acceptance should be taken seriously but how an account of rational acceptance and rational probability judgment can be developed in a coherent and adequate manner.

Case Western Reserve University

BIBLIOGRAPHY

[1] Rudolph Carnap, 'The Aim of Inductive Logic', in *Logic, Methodology and Philosophy of Science* (ed. by E. Nagel, P. Suppes, and A. Tarski), Stanford 1962, pp. 303–18.
[2] Richard Jeffrey, *The Logic of Decision*, New York 1965.
[3] Richard Jeffrey, 'Probable Knowledge', in *The Problem of Inductive Logic* (ed. by I. Lakatos), Amsterdam 1968, pp. 166–80.
[4] P. A. Samuelson, *Foundations of Economic Analysis*, Cambridge, Mass., 1947.
[5] L. S. Savage, *Foundations of Statistics*, New York 1954.
[6] Patrick Suppes, 'Concept Formation and Bayesian Decisions', in *Aspects of Inductive Logic* (ed. by J. Hintikka and P. Suppes), Amsterdam 1966, pp. 21–45.
[7] Patrick Suppes, 'Probabilistic Inference and The Concept of Total Evidence', in *Aspects of Inductive Logic* (ed. by J. Hintikka and P. Suppes), Amsterdam 1966, pp. 64–5.
[8] Patrick Suppes, 'Rational Changes of Belief', in *The Problem of Inductive Logic* (ed. by I. Lakatos), Amsterdam 1968, pp. 186–8.

REFERENCES

* This essay was written with partial support from the National Science Foundation, Grant No. GS-2050.

[1] R. Carnap [1].

[2] *Ibid.*, p. 309.

[3] P. Suppes [7].

[4] In addition to the paper mentioned see [6] as well as [8], which is Suppes' comment on R. C. Jeffrey's paper [3]. Jeffrey's reply is instructive.

[5] See P. A. Samuelson [4], p. 257, and footnote 8, p. 263.

[6] Whether strict coherence is plausible or whether more coherence should be required is debatable. However, adoption of the weaker assumption would complicate the exposition by requiring qualification of the claim that the set of sentences accepted as evidence is identical with the set of sentences accorded a credence value of 1. No point essential to this discussion would be lost, however, if the qualification were to be made.

[7] See, for example, L. S. Savage [5], pp. 46–55.

[8] Patrick Suppes has emphasized this point. See [6], p. 35.

[9] Of course, X might be mistaken in reporting that $c(H, X\text{'s total evidence}) = r$. Indeed, the metalinguistic predicate $c(H, \text{—}\ ) = r$ may be true of no non empty set of sentences in X's language. Observe, however, that X can be mistaken in what he takes to be his explicitly given evidence. The shift from explicit to implicit characterizations of evidence involves no change in this respect.

[10] See, for example, R. C. Jeffrey [2], pp. 156–7.

[11] *Op. cit.*, Ch. 11. Also [3].

[12] Jeffrey [3].

[13] *Op. cit.*, p. 179.

[14] *Ibid.*, p. 176.

RICHARD C. JEFFREY

DRACULA MEETS WOLFMAN: ACCEPTANCE VS. PARTIAL BELIEF

One of the things I'd like to see come out of this conference is a clarification of the issues between Isaac Levi and me.[1] I take it that Levi's scruples about partial belief and probability kinematics are not idiosyncratic, nor are my scruples about his work on acceptance, so that the matter may be of general interest. When Dracula meets Wolfman in the movies it is not simply I-and-thou: They gibber and slaver for all vampires and all werewolves everywhere. So let it be with us.

I. MY KIND OF PRAGMATISM

Levi and I both seem to be pragmatists of sorts – of different sorts. Thus, he quotes with approval from Peirce ([11], p. 208), "The settlement of opinion is the sole end of inquiry", but my kind of pragmatism has deliberation, not opinion, as its focus: It tries to see how much mileage it can get out of the fact that man is an agent. Opinion and valuation come together in the matrix of deliberation, from which action emerges.

I do not mean to deny that men value understanding, and seek it for its own sake as well as for its practical value; but I do not see understanding as a matter of relief from agnosticism, nor do I find it illuminating to characterize scientists as seeking to replace doubt by true belief. (Still less do I find it illuminating to consider such questions as, 'Am I a *true* pragmatist?' and 'Have I betrayed fallibilism?' – see [11], p. 209.)

Deliberation can take a variety of forms, and has done, in the history of mankind. Our notion of what counts as rational decision-making keeps evolving – anyway, changing. The view that I find most satisfactory is very dimly adumbrated in the *Protagoras* (ca. 356) and Aristotle's *De Anima* (434ᵃ 7–10); more sharply in *The Port-Royal Logic* (last Chapter), and more sharply still in Daniel Bernoulli's *Exposition of a New Theory on the Measurement of Risk* (1738); and has finally come into its own in the past 30 or 40 years through the work of Ramsey, De Finetti, Savage, and others. As an important social force, it has come into its own only in the last decade. L. J. Savage's *The Foundations of Statistics* was a radical book, when it appeared in 1954 – it took some years for his 'Personalistic' standpoint to be recognized as a serious and important one – but now the Ramsey-Savage view of decision-making is much in vogue, in the corridors of governmental and economic power. Students at the Harvard Business School learn decision theory from a book [12] by Howard Riaffa, who is Frank P. Ramsey Professor of Managerial Economics and a member of the Graduate School of Public Administration. According to *The New York Times Magazine* ([16], p. 39)

> Decision theory, as it is taught – and, to some extent, created – at the business school, rests on the concept of "subjective probabilities", which allows for the decision-maker's attitude toward risk. Through a device called "the preference curve" (which assesses, say, how much it stings a man to lose $50000 on a chance of gaining $275000) this attitude may be translated into numbers and incorporated into decision-tree calculations.

I have no proof that my own account of rational deliberation is right, or uniquely right.[2] Rather, that account aims at being a fair representation of what currently passes as sensible deliberation. Mind you, part of the reason why what passes, passes is that it fits into such coherent frameworks as Ramsey and his successors have given, for deliberation – we like to feel we know what we are doing, or why we are doing it, in that sense. I am pretty happy with (say) my account in the sense that I think that something like it will continue for a while to serve as an adequate-seeming framework. On the other hand, I am pretty sure that it is not quite satisfactory as it stands; I expect a more satisfactory account to appear soon. But you see what I mean: A currently satisfactory account of practical reason is satisfactory in being a coherent description, not of all practice, but of the instances of practice now thought patently sound – and an account succeeds as a description partly because of its persuasive force.

Perhaps Savage would put the matter more strongly. In his book he presents seven postulates which he hopes everyone will agree are true of all sensible preference rankings of possible acts – everyone: Personalists and those who have never heard of Personalism and those who have heard of it but reject it. He then gives a mathematical proof that anyone whose preference ranking satisfies the postulates is already, in effect, a Personalist, i.e. for any such person there is a probability measure p and a utility function u – where p is determined uniquely by the preference ranking, and u is determined uniquely except for choice of a zero and a unit –, which characterize the ranking. In particular, one act will be ranked with, above, or below another accordingly as its expected utility as computed via p and u is equal to, greater than, or less than the expected utility of the other.

One might find Savage's argument interesting and valuable, but less than compelling. Someone who sees himself as believing in various special decision methods that are incompatible with Personalism and who also sees himself as accepting Savage's postulates as norms will see that he has an inconsistent view of himself, after he follows Savage's proof. Likely as not, he will then have a hard look at the individually plausible-seeming postulates, in which so much proves to be buried when they are viewed collectively; and he may find that certain of them are unacceptable after all. This is a situation of the well-known sort in which you take yourself to believe p, q, and $-r$, and then discover that p and q together imply r. The logical alarm has rung: you will want to give up one (or more) of $p, q, -r$, but logic does not tell you which. By the way, the account of decision-making that I espouse is like deductive logic in that it does not tell you what to do, or what to prefer to what; but it *does* ring a quasi-logical alarm e.g. when you find that you prefer p to q, and q to $p \vee q$, and then discover that p and q are logically incompatible; for in such a case, $p \vee q$ must be somewhere between p and q in your preference ranking, if the 'logic of decision' is not to be violated.

Ethan Bolker[3] has given a set of postulates that bear the same relation to my account of probabilities and utilities that Savage's postulates bear to his. Of course, I find Bolker's postulates acceptable, individually and *in toto*; but I, like Bolker, find that in the case of one of the postulates, I must translate it into terms of probabilities and utilities in order to see that I accept it. But *that* does not discourage me. The conclusion of a valid inference is always lurking somewhere in the premises, and to the extent that the conclusion is a surprise, one has not fully understood the (implications of the) premises! Furthermore, the conclusion is generally weaker than the conjunction of the premises, and therefore more likely to be true; then *prima facie* it seems a bit unpromising to try to convince someone of the conclusion indirectly, by convincing him that the premises are all true, and the inference valid. In the inference under discussion, from Bolker's postulates, I find the conclusion eminently reasonable isimply because the probability/utility way of looking at decision problems has succeeded so well in making sense of a wide diversity of kinds of decisions, some of which were initially rather puzzling. As far as I know, this Personalistic or 'Bayesian' account is unique in its ability to accommodate the whole range of decision problems, while clearly marking out the domains of applicability of various other, more special accounts, e.g. that in which one seeks to minimize maximum disutility. I am a Bayesian because I see the matter in this way – not conversely.[4]

But what has any of this to do with belief? Well, the Bayesian thesis might be summarized by the statement that when A and B are propositions you are sure do not both hold, i.e., when your belief function p is such that $p(A \& B)=0$, and when $p(A \vee B)$ is not 0, then the utility $u(A \vee B)$ which you ascribe to one or the other of A, B's holding ought to be a weighted average of the utilities you ascribe to them separately:

$$u(A \vee B) = w_1\, u(A) + w_2\, u(B),$$

where the weights w_1, w_2 add up to 1 and are proportional to your degrees of belief in A and in B. Then we must have $w_1 = p(A)/p(A \vee B)$ and $w_2 = p(B)/p(A \vee B)$. Here, the promise of $A \vee B$ is viewed as a gamble with two possible outcomes: A, B. Your valuation of $A \vee B$ is then determined by your valuations of A and of B, and by your beliefs about A and about B, where 'belief' here is understood as an attribute of your attitude toward various risks. In particular, in terms of your attitudes toward various gambles in which only small gains and losses are possible (so that your utility curve is fairly linear) your degree of belief in a proposition A is p if and only if you think it fair to bet on A at odds of $1-p:p$ (e.g., to pay \$$p$ to get \$1 if A is true, and lose your \$$p$ if A is false).

It is one of the charms of the Bayesian position that it uses a notion of partial belief which is clearly and simply related to the agent's attitudes toward risk. The charm is twofold. For one thing, it gives one fairly clear, quasi-operational criteria for determining whether someone has a definite degree of belief in a certain proposition at a certain time and, if he has, for discovering what that degree of belief is; and it gives one a rather beautiful way of showing that degrees of belief ought to satisfy the axioms of the elementary probability calculus, viz., that degree of belief ought to be a normalized, non-negative, additive function of propositions.[5] This notion of degree of belief accords well enough with one strand in our talk about belief, but it is not intended as a bit of ordinary language analysis or as a bit of phenomenology. This notion of belief is the theorist's, and need not be the agent's. Thus, suppose the agent insists that he believes that Black Beauty will win the first race at Suffolk Downs – insists that he fully believes this, is certain of it, etc.; and suppose that he will not give any longer odds than 9:1 on Black Beauty, but *will give* odds of 9:1 or less. I would conclude that his degree of belief in Black Beauty's winning is .9; I would also conclude that he is using the terms 'full belief' and 'certainty' in ways that elude me. But no matter – I shall give him the word 'belief', if he insists, and use 'degree-of-belief' in the clear sense described above, as a technical term.

It need not be easy to tell whether someone *has* a degree of belief in a certain proposition at a certain time, e.g. his attitude toward risks in which that proposition figures might be a rather complex one. One mode of complexity that is compatible with the Bayesian attitude was studied by Friedman and Savage [5]: The bettor's utility function for money may be such that the magnitudes of the possible monetary gains and losses, and not just the odds, *gain*:*loss*, figure in his decision to accept or reject bets. If this is the case, we shall be able to find it out by examining his attitude toward various gambles and certainties, and rationalize his behavior by observing that although the monetary odds may be the same in two bets, the utility odds are different. Thus, an even-money bet for \$100 on the toss of a fair coin might be unattractive because the utility gain

(a) $u(\text{being }\$100\text{ richer}) - u(\text{status quo})$

is less than the utility loss

(b) $u(\text{status quo}) - u(\text{being }\$100\text{ poorer})$

so that the utility odds, a:b, are poor although the money odds 100:100 are even. Concretely, plotting the utilities of various incomes against the dollar sizes of those incomes as in Figure 1, the expected utility of a fair, even-money bet for \$100 is represented by the height of the midpoint M_1 of the dotted line segment connecting the points L and G_1 which represent the two possible outcomes: lose \$100 and gain \$100. But the sure income of which the utility is equal to the expected utility of the bet is $-\$50$: Having made the bet, but not yet knowing the outcome, the bettor whose utility function is as shown in Figure 1 would be willing to pay up to $50 to be free of the situation – so much (according to Figure 1) does his dread of losing outweigh his hope of winning, even though he thinks himself as likely to win as lose. In general, where the utility function is concave downward, the agent is adverse to risk. But where the stakes are different, the same bettor may find risk attractive, as in a situation where he will be $100 poorer if a fair coin falls tail up, and $300 richer if it falls head up. Again, the expected utility of the gamble is represented by the height of the midpoint M_2 of the dotted line segment connecting the points L (lose $100) and G_2 (gain $300). The sure income of which the

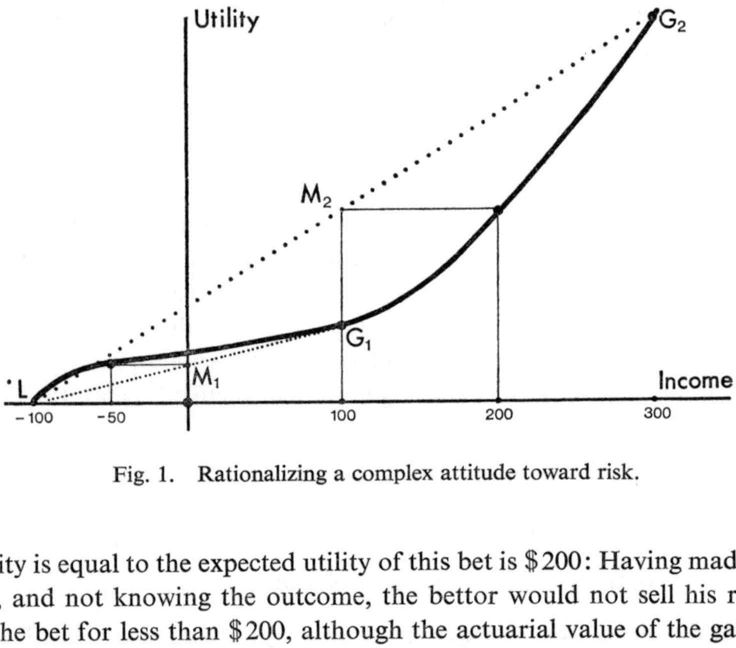

Fig. 1. Rationalizing a complex attitude toward risk.

utility is equal to the expected utility of this bet is $200: Having made the bet, and not knowing the outcome, the bettor would not sell his rights in the bet for less than $200, although the actuarial value of the gamble is only $100. Then apparently clashing attitudes toward risk can make perfect sense, if the agent's utility function has the right sorts of non-linearity. The particular sort of nonlinearity shown in Figure 1 would be appropriate to someone who could be $50 richer or poorer without radical change in his situation, but for whom greater gains or losses would appear disproportionately attractive or repulsive – greatly expanding or contracting his horizons. (Perhaps with another $200 he could invest in a very profitable enterprise, while with $200 less he would be bankrupt.)

In other cases, especially where the payoffs are not monetary, the truth or falsehood of the proposition gambled upon may influence the utility of winning or losing in a confusing way. Thus, I may think the weather tomorrow as likely to be foul as not and, quite apart from that, I might rate a ticket to a football game tomorrow as high above the *status quo* as I rate a dull day around the house below it. Abstractly, then, a bet in which I get the ticket if I win and the boredom if I lose ought to look fair to me if I have a degree of belief 1/2 in the proposition bet upon; yet, the bet would be unattractive to me if I were betting on foul weather tomorrow, since that proposition would detract from the utility of winning, while its denial would leave the utility of losing as it was. (See Figure 2.)

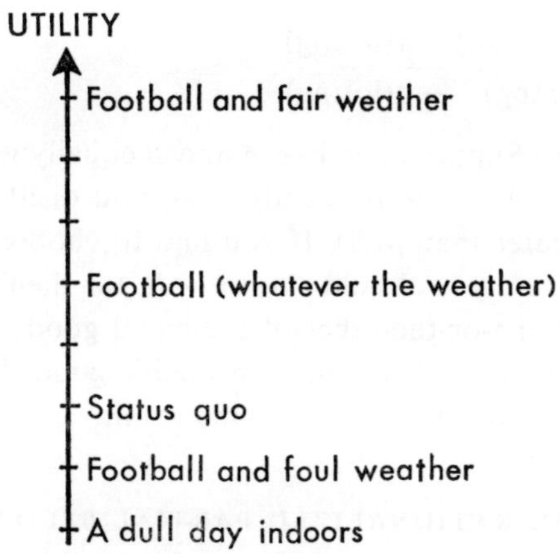

Fig. 2. Enhancement.

To determine whether $p(A)$ exists, for a certain agent, we may have to go far afield, and examine his attitude toward risks in which propositions other than A figure. Imputations of beliefs and values to agents face the tribunal of the agent's attitudes toward risk as a body, although some imputations are especially easy to isolate and test.

The intimacy of the connection between valuation and belief is very evident when we let $B = -A$ in the relationship

$$u(A \vee B) = \frac{p(A)}{p(A \vee B)} u(A) + \frac{p(B)}{p(A \vee B)} u(B),$$

which holds when A is logically incompatible with B and $p(A \vee B)$ is not zero. Since $p(A \vee -A) = 1$ and $p(-A) = 1 - p(A)$, we then have an equation which can be solved for $p(A)$ to get

$$(*) \qquad p(A) = \frac{u(T) - u(-A)}{u(A) - u(-A)}$$

as long as $u(A) \neq u(-A)$. Here, T is the necessary proposition: $T = A \vee -A = B \vee -B = C \vee -C$, etc. Intuitively, T or $A \vee -A$ is a gamble on the proposition A in which the agent gets A if he wins and $-A$ if he loses. In this as in other cases, the gamble is fair (by the agent's lights) if and only if he takes his probability of winning to be

$$\frac{u(\textit{status quo}) - u(\text{losing})}{u(\text{winning}) - u(\text{losing})}.$$

A final illustration: Suppose you like A and B equally well; does it follow that you dislike $-A$ and $-B$ equally? No, you dislike $-A$ more than $-B$ if $p(A)$ is greater than $p(B)$. If you had to choose between $-A$ for sure and $-B$ for sure you should choose $-B$, for then you have a better chance of getting one-or-the-other of the equal goods A, B, than if you had chosen the denial of A, the more probable good. If you work it out algebraically via (*) you will get the same result.

II. RATIONALIZED PARTIAL BELIEF

It seems pretty clear that conformity with the laws of the elementary probability calculus is a necessary condition for reasonableness of a set of partial beliefs, and it should be equally clear that the condition is not sufficient. But let us have a closer look at both claims, especially the second.

Doubts about the first claim have been expressed by Suppes [15] and Hacking [6]. According to Suppes,

> A theory of rationality that does not take account of the specific human powers and limitations of attention, memory and conceptualization may have interesting things to say, but not about human rationality.

Thus, according to the elementary probability calculus, we must have $p(A) \leqslant p(B)$ whenever we have $A \vdash B$ – where $\vdash$ is the deducibility relation in some complete formalization of first-order logic, say. But since there is no general decision procedure for the relation $\vdash$, it seems inappropriate to count someone as irrational whenever $p(A)$ and $p(B)$ both exist and $p(A) > p(B)$ but $A \vdash B$. Better, indeed, to count his degrees of belief as incoherent, and add that with the best will and the best brain in the world, the poor fellow cannot be expected to be aware of all entailments. Of course that is true, and of course we were not thinking of locking him up as a mental incompetent. The point is merely that whether he knows it or not, his beliefs suffer from a logical failing. We would count *him* irrational only if it had been demonstrated to him that $A \vdash B$ and $p(A) > p(B)$, but he regarded the situation with equanimity. And presumably the fellow, being rational, will not want decision theory to be so permissive as to neglect to classify the situation we have been envisaging as a fault, on the ground that it was not *his* fault.

As to the second claim, that conformity with the laws of the elementary probability calculus is not sufficient for rationality of a set of beliefs – I take this to be clearly correct. I mention this fact because Isaac Levi seems to base some of his arguments against my position on the premise that I think conformity with the laws of the probability calculus is sufficient as well as necessary for reasonableness. This passage is a case in point:

> Thus, if coherence is the only obligation imposed on a rational agent in assigning probabilities to propositions, acquisition of new evidence dictates very little to him regarding how he is to revise his probabilities. New evidence is virtually irrelevant to the revision of probabilities ([11], p. 198).

I conjecture that Levi is reasoning as follows: "The only conditions Jeffrey states, as necessary for rationality, are those of coherence, i.e., conformity with the laws of the elementary probability calculus. If he thought there were any further necessary conditions, he would surely have stated them. Since he has not, it must be that the conditions he has stated as necessary are collectively sufficient, in his view."

What I really think is this: Coherence is far from sufficient for rationality, but I do not know how to state any further conditions, except in fuzzy or circular or inaccurate ways; so I have stated no further conditions. Yet, I am sure that further conditions exist. It is a bit as if I were writing an instruction manual for bicycle riding, and gave a number of suggestions, e.g. 'If you feel yourself falling to the left (right) while moving forward, turn the front wheel a bit to the left (right)', and other such tips. I am quite aware that conformity to all suggestions will not make someone a competent bicycle rider, and I would make further suggestions which would complete the set of tips if I knew how; but I do not.

And the situation is a bit like cycling in another respect, I think. Bicycles are made for man. People generally *are* able to learn to ride them, on the basis of such tips as I have mentioned, after some practice. Our nervous and skeletal systems and musculature and sensoria have various special features, common to men and different from the analogous features in dogs, mackerel, etc., and in virtue of some of these features we are in fact able to learn to ride bicycles even though we are unable to give complete explanations of how we do it. Similarly, I think, for the business of forming our systems of partial beliefs. Being the sorts of creatures we are, we are sharply limited (although in ways I cannot clearly describe) in our capabilities for partial belief. And when we add the requirement that our partial beliefs form a coherent probabilistic structure, the limitations become even sharper, although I cannot give a tight general account of how, or why. But I can give some examples:

The Chevalier de Méré. In a letter to Fermat dated July 29, 1654 (see [17], p. 11), Pascal mentions

> a difficult point which astonished M. so greatly, ...
> The odds on throwing a six with a die in 4 trials are 671 to 625.
> But the odds are *against* throwing a double six with a pair of dice in 24 trials. Nevertheless, 24 is to 36 (which is the number of faces of two dice) as 4 is to 6 (which is the number of faces of one die).

We do not quite know the background, but it may have been this: The Chevalier had been betting on double six in sequences of twenty-four tosses of a pair of dice, and losing, on the whole, even after many changes of dice. Pascal pointed out to him that if (as he said he did) Méré regarded all 36 possible outcomes of a single throw of a pair of dice as equally probable, and if he regarded successive throws as independent (as he said he did) then he must regard two sixes on twenty-four throws of a pair of dice as rather less probable than one six on four throws of a single die, and even as a bit less probable than having an even number of points turn up on one throw of one die. In fact, on the stated assumptions, we must have $p(\textit{one six on four tosses of a die}) = 671/1296 > 1/2$, and

p(*two sixes on 24 tosses of a pair of dice*) $= [36^{24}-35^{24}]/36^{24} < 1/2$, despite the misleading symmetry, viz., 24:36::4:6. In the face of all this, the Chevalier *might* drop the assumption that all 36^{24} possible outcomes of 24 tosses of a pair of dice are equally likely; but being human, he was influenced by the observed frequencies, which conform better with the attitude that p(*two sixes on 24 tosses*) $< 1/2$ than with his original attitude. Still, according to Pascal ([17], p. 11) the discrepancy

> was his great scandal, which made him say haughtily that the theorems were not consistent and that arithmetic was demented. But you will easily see the reason by the principles you have.

These principles were rather arcane, in 1654, but today they are the common property of millions of high school seniors.

The Birthday Problem [1]. Would you accept an even-money bet for a dollar on at least two people in this room[6] having the same birthday? Perhaps some of you would, and some would not, and some of the discrepancy might be accounted for by different nonlinearities in your utility functions for income. But there might remain a basic disagreement, evinced perhaps in the advice you would give to someone whose utility curve for money was linear in the interval $\pm\$1$ about his *status quo*: some of you would call it a good bet, others a bad one, perhaps. The fact is that it is a good bet if there are at least 23 people in the room. When I make this plonking statement, I am making use of facts we all have, about human births, to say that if any one of you will think the thing out in the light of those facts, you will find that your degree of belief in there being two or more people with the same birthday in this room is greater than 1/2 if you count and find that there are 23 or more people here. The facts are, that there is no reason to suppose that the manner of our selection and self-selection for presence here was biased against sameness of birthdays; that the period about 9 months after the Christmas season is one in which more people in this country have their birthdays than other periods of the same length in the year; and the like. Then if we suppose that each of us is as likely to have his birthday on one day of the year as on any other, and that the probability of one of us having a certain birthday, given that some other of us has some (same or other) definite birthday, is the same as the absolute probability of the first person's having the birthday in question, we shall be making a set of assumptions on which, if anything, it is less likely that two of us have the same birthday than is really the case, according to our more accurately articulated belief functions. And on those assumptions, the probability is a bit over .5 that with 23 people here, two or more of us have the same birthday, while the probability is a bit under .5 if there are 22 people here.

Here, then, are two examples which illustrate how a determination to make one's beliefs conform to the calculus of probabilities can result in what will generally be regarded as a correction of one's original belief system. First, a factual or logical alarm rings: One's beliefs are at odds with observed frequencies (factual alarm), or with each other, if one accepts the laws of probability (logical alarm). In neither case can one say quite generally how we think we should respond after due consideration. One may finally regard a discrepancy between observed frequencies and the frequencies one takes to be most likely as a matter of chance, and for the time, anyway, keep to one's belief function, as when a coin is tossed ten times and there are seven heads; this is rather unlikely, but not astonishing, on the usual views about coin-tossing; and it may be that the coin appears quite normal, so that one has no inclination to adopt an extraordinary view of what is going on. Similarly when the logical alarm rings, it is not generally clear how one will think one should revise one's beliefs; but in practice there is a striking uniformity of response to similar situations. Here's a final example.

The Maturity of Chances. How do you go about convincing someone that he is wrong in thinking that, e.g., with a normal coin, the sequence *hhh* is more likely to continue with a *t* (tail) than with an *h* (head)? Sometimes, you cannot. I mean, sometimes you cannot produce reasons or data that will make him adopt the belief function which assigns probability 1/16 to each possible string of four *h*'s and *t*'s. But usually, such people do not have an internally coherent view of what goes on in coin-tossing according to which $p(hhht/hhh)$ is, say, greater than 1/2. If such a person will agree that any set of degrees of belief in such propositions ought to be extendible to a set which is coherent and assigns degrees of belief to each of *hhhh*, *hhht*, *hhth*, ..., *tttt*, you can generally rely on any particular such extension's having features which disturb him in a way that overbalances his satisfaction at having $p(hhht/hhh)$ be as he originally had it. You might try tossing coins with him, and showing him that among the runs of three heads, about the same number are followed by tails as by heads, but that need not work, partly because you have to toss a large number of times to accumulate much data of that sort, and partly because he might have $p(hhht/hhh)$ only very slightly above 1/2, so that it might take months to collect data he regards as significant. In that case, of course, the two of you might agree to disagree, reasoning that there is little or no practical difference between your views.

I have spoken at length about coins and dice because these are cases where there is broad intersubjective agreement – and even there, deductively reasonable men may differ, although they seldom do for long. In this domain it is entirely possible to go strikingly far beyond mere coherence as a statement of what constitutes a sensible belief function, e.g. one might state that concerning an ordinary penny, tossed in an ordinary manner, no belief function is rational unless it attributes nearly identical probabilities to any two equally long strings of *h*'s and *t*'s. The belief functions that have this feature are rational in the sense that there are commonly available empirical data and arguments about coherence which fairly uniformly have been found to convince people who care to think about it that, after all, their beliefs are as described by one of the 'rational' belief functions; and in the sense that such people regard themselves as having good reasons for abandoning their old belief functions and adopting one of the 'rational' ones.

Then as a practical matter, I think one *can* give necessary conditions for reasonableness of a set of partial beliefs that go beyond mere coherence – in special cases. The result is a patch-work quilt, where the patches have frayed edges, and there are large gaps where we lack patches altogether. It is not the sort of seamless garment philosophers like to wear; but (we ragged pragmatists say), the philosophers are naked! Indeed we have no proof that no more elegant garb than our rags is available, or ever will be, but we have not seen any, yet, as far as we know. We will be the first to snatch it off the racks, when the shipments come in. But perhaps they never will. Anyway, for the time being, we are dressed in rags, tied neatly at the waist with a beautiful cord – probabilistic coherence. (It is only the cord that visibly distinguishes us from the benighted masses.)

What might the seamless garment look like? Well, Carnap has an idea: It will be a conditional probability measure $c(\ /\ )$ defined on a unified language of science which we shall also use in daily life. In that language we shall be able to report the weather, meter readings, and upset stomachs as well as the laws of physics-chemistry-biology as we shall then take them to be. For any sentences p and q in that language, $c(p/q)$ will have a definite value (unless $c(q, p \vee -p)$ is zero) which will presumably be computable as accurately as we please, with the aid of superfast, complex machinery then available. 'In principle', we shall then be able to separate the inductive logical component (represented by the function c) from the experiential component (represented by an enormously long conjunction e of observation reports) in anyone's beliefs, as they ought to be. Of course, there are difficulties, e.g. to accumulate the enormous conjunction e such that for any hypothesis h, $c(h, e)$ is the degree of belief in h that is justified by one's experiences, one would have to spend all one's time writing in one's diary, and one would consequently have little of interest to write; and there are other difficulties, connected with experiences one cannot clearly formulate in the language. But, Carnap would say, the philosophical point is that in principle we then have an empiricist account of justified partial belief, whatever the difficulties of carrying it out in practice.

Indeed, at such a stage in the development of mankind, practice might well be much different from what it is now, e.g. the Academy of Unified Science might be constantly feeding certified evidence-sentences into the machine, to go into a common pool of data. Individuals might also transmit other data for storage in their personal portions of the great Memory Banks via their pocket transceivers, on which they could also request and receive up-to-date rational (for them) degrees of belief in hypotheses that interest them at the moment. It might even be possible to wire transceiving equipment into people's heads, to do much of this automatically. Perhaps in the end we should then have evolved into a colonial form of life, where to a great extent we share a supersensorium and a superconsciousness – one for all.

That is as may be – it is not my object to predict how the race will develop in the long run, if there is one. But as far as I can see now, the c-function built into that supercomputer will be the (or *a*) rationally correct one in at best the evolutionary sense that, when the computer is built, it is the survivor (or one of the survivors) in the process of proposing new candidates for the position and eliminating some of the contestants as having features that almost nobody finds acceptable. In fact, *this* process is a far more tenuous one than the process illustrated above in the case of the Chevalier de Méré etc., because in this process we have no factual alarms – only logical ones. The value of $c(s_1/s_2)$ is what it is independently of what we have observed; that is what makes c a purely 'logical' function, so that the entire experiential component in our rational belief is represented by the particular choice of $s_2 = e =$ one's diary to date.

Then here is where I stand, at the moment. The human race is evolving culturally as well as physically, and part of what is evolving culturally is our notion of what constitutes sensible decision-making and what constitutes rational belief, where by 'belief' I mean the thing that goes along with valuation in decision-making: degree-of-belief, or subjective probability, or personal probability, or grade of credence. I do not care what you call it because I can tell you what it is, and how to measure it, within limits, no matter what you want to call it. (On the other hand I think it an instructive comment on the way we normally talk about belief, to call it 'degree of belief' – but that is secondary. The analysis of ordinary language is not my object either.)

God knows, we may be wrong in having the view we have, of rationality in these matters. To demonstrate this would be to give a persuasive analysis of what constitutes right and wrong in this context, and then show that our current view (to the extent that there is a single current view) is wrong. Better yet, one would provide a new view, that is demonstrably right! But if and when this happens, it will (as far as I can see) be just another instance of cultural evolution: The persuasive analysis will have persuaded the best minds (let us call them) to change their minds, and the Harvard Business School will follow, and government and industry will follow them. And then we shall be in a position rather like the one we are in now: Relative to what will then be called 'now', *as far as we can see now, such-and such is the best account of deliberation and rational belief.* There will, of course, be this difference: at that time we shall presumably have a more copious account, within which we can see the limits and the limited virtues and defects of the present account. That will no doubt be a large ingredient in the confidence men will then feel, in keeping to their account instead of regressing to ours.

Nor would I bet very heavily on our notions of utility and degree of belief being usefully discernible in men's view of rational deliberation, a century from now, except for certain vestiges; nor am I disturbed by the fact that our ordinary notion of *belief* is only vestigially present in the notion of degree of belief. I am inclined to think Ramsey sucked the marrow out of the ordinary notion, and used it to nourish a more adequate view. But maybe there is more there, of value. I hope so. Show me; I have not seen it at all clearly, but it may be there for all that.

III. ON THE DYNAMICS OF PARTIAL BELIEF

In Physics, Dynamics is a contrary of Kinematics as well as of Statics: it is the first contrariety that I had in mind when I called Chapter 11 of *The Logic of Decision*, 'Probability Kinematics'. Take a see-saw, with fulcrum 2/3 of the way toward your end. If you push your end down two feet, the other end will go up three. That is kinematics: You talk about the propagation of motions throughout a system in terms of such constraints as rigidity and manner of linkage. It is the physics of position and time, in terms of which you can talk about velocity and acceleration, but not about force and mass. When you talk about forces – *causes* of accelerations – you are in the realm of dynamics.

So the kinematics of partial belief is concerned with the question, 'Suppose you have a definite coherent set of degrees of belief in various propositions, and then your degree of belief in one or more of them changes; how must that change be propagated throughout the entire system of your beliefs, if coherence is to be preserved?' There is no general answer, any more than there is a general answer to the question, 'Suppose point *A* rises, on the rigid bar shown in Figure 3. How will point *C* move?'

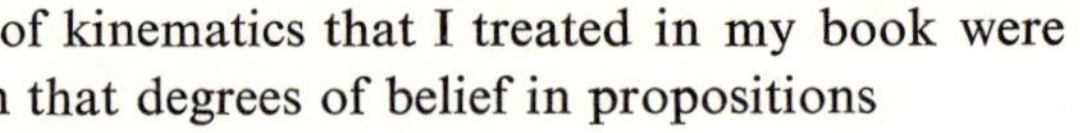

Fig. 3.

You need to know a bit more than that, in order to get an answer, e.g. it would be enough to know that point *B* remains stationary.

The cases of kinematics that I treated in my book were of this sort: We are given that degrees of belief in propositions

$$(\text{'Basis'})\quad A_1, A_2, \ldots, A_n$$

change from their present values, $p_0(A_1)$, $p_0(A_2), \ldots, p_0(A_n)$, to new values, $p_1(A_1)$, $p_1(A_2), \ldots, p_1(A_n)$; we are also given that the A's form a partitioning, i.e., for all $i, j = 1, 2, \ldots, n$,

$$\text{(Partitioning)} \quad \begin{cases} p_1(A_i) \neq 0, \\ p_1(A_i \,\&\, A_j) = 0 \text{ if } i \neq j, \\ p_1(A_1 \vee A_2 \vee \cdots \vee A_n) = 1. \end{cases}$$

Finally, we are given that all conditional probabilities, relative to the individual A's, are the same after the change as they were before: For all $i = 1, 2, \ldots, n$, and for all B for which $p_0(B)$ exists, we have

$$\text{(Rigidity)} \quad p_1(B/A_i) = p_0(B/A_i).$$

If all these conditions hold, and if both p_0 and p_1 satisfy the laws of the elementary probability calculus, then the function p_1 is determined: For any proposition B for which $p_0(B)$ exists, we have ('K' for 'Kinematics')

$$\text{(K)} \qquad p_1(B) = p_1(A_1)\, p_0(B/A_1) + \cdots + p_1(A_n)\, p_0(B/A_n).$$

The proof of all this is as trivial as the proof of Bayes' theorem. It is enough to verify that by the elementary probability calculus, (Partitioning) implies

$$p_1(B) = p_1(A_1)\, p_1(B/A_1) + \cdots + p_1(A_n)\, p_1(B/A_n),$$

and then substitute via the rigidity conditions, to get (K).

An obvious sort of application is that in which some passage of experience leads the agent to change his degrees of belief in the A's, but does not lead him to change his conditional degree of belief in any proposition B relative to any of the A's. Then the rigidity conditions hold, and if the A's form a partitioning and his belief function is coherent both before and after the change, his new belief function is determined: It is as in (K). On this, Levi [11] comments that coherence is a static condition, which may be met by the agent's beliefs both before and after the passage of experience even though (K) fails: Both belief functions may be coherent, and we may indeed have $p_1(A_1), \ldots, p_1(A_n)$ as his new degrees of belief in the A's which form a partitioning, but nothing in the coherence conditions forces the function p_1 to be as in (K). (I *think* that is what he is saying in the first part of part II of [11].) Now that is true enough, *but if the rigidity conditions also hold* (as they must, if K is to be applicable), *then it is a matter of logic and high-school algebra that K holds too.*

Starting on p. 199, Levi [11] considers a situation in which the agent is sure that a certain urn, from which he draws a disc, is composed of blue discs together, perhaps, with some green ones. Let A_1 be the proposition that the disc drawn is blue, and A_2 that it is green. These two A's form a partitioning. Now Levi considers three cases:

(1) The agent fails to observe the disc, but somehow or other (perhaps because of a change in his blood chemistry) comes to have new degrees of belief $p_1(A_1)=.9$, $p_1(A_2)=.1$ in the A's anyway.

(2) The agent observes the disc 'and in response admits... [A_1]... into his evidence'. I take it that this implies a change in the agent's degree of belief in A_1 to a new value of $p_2(A_1)=1$, although I am not sure. 'Acceptance as evidence' is a technical term in Levi's account ([10], pp. 28–9) of rational belief.

(3) The agent observes the disc (presumably, under less than ideal lighting conditions, or through sunglasses) and in response his degrees of belief in the A's change to new values, $p_3(A_1)=.9$, $p_3(A_2)=.1$.

Now let's consider these cases in turn and, while we are at it, let us consider what Levi says about them.

Case 1: The agent does not observe the disc, but for some reason comes to have new degrees of belief in the A's anyway. His new belief function is p_1, which reflects these changes and perhaps others as well. (Some other changes will be required if p_1 is to be coherent, unless the set of propositions about which the agent has beliefs is very simple indeed.) *Questions:* (a) Are the agent's new judgements, represented by p_1, rational? (b) Was the shift from p_0 to p_1 rational? (c) Was it irrational *of the agent* to change from p_0 to p_1? Levi distinguishes (b) from (c) in a way that I do not: As far as I can see, rationality of a set of beliefs or of a change in beliefs is always relative to an agent at a time – unless you can manage to give a thorough description of the mental state and the experiences which, in that or any other agent, support a verdict of 'rational' or 'irrational' on the beliefs or on the change. Certainly, I see no hope of judging from an examination of p_0 by itself whether it is a rational set of partial beliefs, or of judging from an examination of p_0 and p_1 whether a shift from p_0 to p_1 is a rational one! I still do not want to call the agent himself irrational when, e.g., his belief function p is such that $p(A)>p(B)$ even though one can show by a complicated proof that $A \vdash B$. In that case I would call p irrational even though the agent may be highly rational, as human beings go. Similarly when the agent's belief function p is coherent, but reflects his missing of some abstruse sensory cue that is available to him, I would not call the agent irrational, but would want to make some deprecatory remark about p as a belief function for him at that time, or perhaps as a belief function for anybody who is in the relevant respects as the agent is at that time. The deprecatory remark need not contain the word 'irrational'. It would be more helpful to be more precise, e.g. to remark that the agent seems to have missed a certain sensory clue; and similarly in the first case, it would be more helpful to point out that, unbeknownst to the agent, his belief function p is incoherent, than to give the less informative description, 'It is irrational.' And in the case of the missed sensory clue, it would probably be misleading to characterize the situation even vaguely as one in which p is *irrational* for the agent at that time; 'insensitive' might be a better term. My general evolutionary view of our changing conceptions of what constitutes sensible decision-making and sensible ways of adjusting one's beliefs to one's experiences makes me chary of talk of 'rationality', as does the weight I assign to tacit elements in the basis for our judgements of such matters.

To get back to case 1: I simply do not know what to say about the agent's shift from p_0 to p_1 in the absence of any information about what caused it. If, as Levi suggests, the cause was a change in his body chemistry quite unrelated to any ratiocination or perception on the agent's part, then I would call the shift mysterious and a-rational. Furthermore, if I were informed that p_0 was quite a sensible belief function for the agent at the time, I would deplore the shift to p_1, and view p_1 as unsatisfactory; I would not even balk, here, at calling p_1 irrational for that agent at that time, and saying that the agent himself acted irrationally.

Case 2: The agent observes that A_1 is true. Then his degrees of belief in the A's change to $p_2(A_1)=1, p_2(A_2)=0$. (Let me discuss this particular reading of Levi; if it is a misleading reading, Levi will correct me.) If this is indeed the whole of what he takes himself to have learned from the observation, then (as far as we can tell from *this* story), nothing has changed his degrees of belief in any propositions conditionally on A_1. We then have $p_2(B/A_1)=p_0(B/A_1)$ and since $p_2(A_1)=1$, the elementary probability calculus gives

$$p_2(B)=p_0(B/A_1)$$

for all B's for which p_1 is defined. (He no longer has any conditonal degrees of belief relative to A_2, i.e., $p_2(B/A_2)=0/0$ – undefined. Alternatively and, by and large, harmlessly and uselessly, we might stipulate that $p_2(B/A_2)=p_0(B/A_2)$.) But maybe there is more to the story. By augmenting the story in various ways, we might make it plausible that p_2 is not as described above. Thus, we might imagine that seeing the disc to be blue sets off or is anyway accompanied by some train of ratiocination which makes him doubt the soundness of his original judgement, that $p_0(B/A_1)=.7$, say, for some particular B. Perhaps he takes $p(B/A_1)=.4$ to be a sounder judgement, and in consequence forms p_2 by a process more messy than conditionalization of p_0 relative to A_1. Then I would say that conditionalization would have been inappropriate after all – just as I would call (K) inappropriate if one of the rigidity conditions failed. This inappropriateness is of the clear-cut kind, where the new belief function would be incoherent; if the chain of ratiocination leads the fellow to the attitude that $p_2(B/A_1)=.4\neq p_0(B/A_1)$ and he forms p_2 by conditionalizing p_0 relative to A_1, then p_2 will not be single-valued! (Clearer, perhaps, to put it this way: the change in conditional probability of B relative to A_1 simply *prevents* him from forming a new, coherent belief function by conditionalizing p_0 relative to A_1.)

This is the sort of thing I find unsatisfactory about Levi's examples: He tells a little story about what the agent has or has not observed, and tells you that he has changed his degrees of belief in the A's in a certain way. Now it is natural to suppose that unless the story calls out for completion (e.g. by 'the change was caused by an attack of indigestion' or 'the change was caused by a brilliant new idea that just happened to come to him then' in case 1) then we have been told everything relevant; for we are asked, on the basis of the story, to say whether the change was a reasonable one – small wonder that we suppose we have been given all the relevant facts. Especially in case 2, when we are told that the agent observes the disc and consequently accepts A_1 as evidence, it is natural to suppose that nothing else, e.g. ratiocination, is going on that might bear on the appropriateness of the simplest change, from p_0 to $p_2=p_0(\ /A_1)$. But then Levi faults that simple change by pointing out that it is not the only one that will eventuate in a coherent function p_2; and indeed, there are plausible stories one can tell when conditionalization relative to A_1 would be inappropriate, e.g. one could continue the story of case 1 by saying, 'and the agent also observed that B is true', whereupon (if *that is* the whole story) we should have $p_2(C)=p_0(C/A_1\,\&B)$ for all C for which p_0 is defined.

The real use Levi makes of this ploy – that the truth need not be the whole truth – is to set a scene in which I seem committed to the truth of the antecedent of this conditional:

if coherence is the only obligation imposed on a rational agent in assigning probabilities to propositions, acquisition of new evidence dictates very little to him regarding how he is to revise his probabilities. New evidence is virtually irrelevant to the revision of probabilities.

But as I have said repeatedly above, I do not take coherence to be sufficient for reasonableness of someone's beliefs. The situation is rather that coherence is the only condition I can think of which is *necessary* for reasonableness of sets of beliefs in all cases – regardless of what the agent's mental and experiential condition may be. In case 1, where the story is clearly incomplete, the only thing I can tell you about the reasonableness of the agent's beliefs without hearing the rest of the story is that if they are incoherent, they are unreasonable. Tell *me* more and maybe I shall be able to tell *you* more; but please do not imagine that because I cite no conditions beyond coherence as necessary in all cases, I think that coherence is sufficient in any case! Now, finally, let us look at

Case 3: The agent observes the disc. He admits no new proposition into his evidence but changes to new degrees of belief $p_3(A_1)=.9$, $p_3(A_2)=.1$ in the A's. Levi says ([11], p. 203)

One might be tempted to suggest that perhaps case (3) is not completely described.

Indeed I am sorely tempted, and herewith yield to the temptation, although not (as Levi notes) by continuing the story, 'These degrees of belief in the A's are simply $p_3(A_1)=p_0(A_1/\text{the chip appears to be blue})$ and $p_3(A_2)=p_0(A_2/\text{the chip appears to be blue})$, where "The chip appears to be blue" is a phenomenological report which describes the relevant aspects of the agent's observation.' My objection to *this* move is that I see nothing in the statement, 'The chip appears to be blue', to warrant $p_0(A_1/\text{the chip appears to be blue})=.9$ instead of .7, say – the statement is too vague for that.

Rather, I would complete the story by saying, 'And nothing in what the agent saw or thought on that occasion moved him to revise his degrees of belief in any proposition relative to either of the A's.' Indeed, if you had to guess the rest of the story, you would guess some such thing simply because you expect the narrator to have said something about it, if in fact something did move the agent to change some of those conditional probabilities. Anyway, with the story completed in that way, p_3 *must* come from p_0 via (K) if both belief functions are to be coherent.

My general point is this: To judge the soundness of a shift from p to p' we must look at more than the kinematics; we must not only look at the two belief functions, and their differences; we must also inquire into the forces which prompted the change – the dynamics of the situation. This is another way of saying that although coherence is the only condition I can formulate that seems necessary for reasonableness of belief functions in all cases, each particular case must be examined with an eye to the agent's particular situation. In particular, the question, whether (K) is an appropriate kinematical relation in a particular case must be answered by finding out what is moving the agent in that case. If he has just examined a chip in sunlight while wearing his sunglasses and he offers 9:1 but no longer odds on the chip's being blue, and wants to *be given* those same odds on its being green, then it is *plausible* to suppose that the rigidity conditions are met, and that (K) is appropriate with .9 and .1 substituted in the right places. Of course, one may be wrong, and one can describe various observations one might make that would convince one of that.

Of course one can, in a rather question-begging way, give necessary and sufficient conditions for appropriateness of (K): it is necessary and sufficient that it be sensible for the partitioning conditions to hold of the agent's original belief function, and that the rigidity conditions hold between his old and new belief functions – in view of whatever the agent's situation happens to be at the time of the change. Similar question-begging conditions can be given for conditionalization relative to E in response to an observation: Everything relevant to the agent's beliefs that can be said about the observation is expressed by the statement, 'The agent has, sensibly, come to have degree of belief 1 in E.' But none of this is much real help. What is of help is our practice, of using conditionalization – or, perhaps, (K) – in various cases more or less tacitly seen to be the right ones. Any *one* of these cases can be discussed non-circularly with some profit – in terms of facts about *that* case which perhaps are generally seen as supporting or undermining the thesis that conditionalization is indeed sensible, there. But the genus of cases in which conditionalization is appropriate is one I do not know how to characterize clearly and noncircularly. Ditto for cases in which (K) is appropriate.

IV. THE AUTHORITY OF REASON

According to Aristotle[7] humans of both sexes are rational animals, but in women, reason is without authority. ('I contradict myself? Very well, I contradict myself.')

In what, if anything, does the authority of reason consist? In particular, does not my evolutionary account of our canons for reasonableness of deliberation and belief tend to undermine the authority of reason? What reason is there to think the current vogue any better than the one before that? None – if you require a reason to be certifiable as such by certifiably eternal canons of rationality that transcend all vogues. But if you are less exacting, the answer is not far to seek: The reasons in favor of the current vogue are the considerations which prompted us to see it as satisfactory, or as more satisfactory than any other canons that were in the running at the time the current vogue swept the field. To inquire into the source of the authority of reason is to ask the question wrong way round: What we currently take to be reason has the status of reason in our eyes because of the authority it has come to have, for us. Granted, we can rebel against the authority of what currently passes for reason; but remember that man is on the whole a reason-seeking animal, shunning frank irrationalism; and remember that not just anything can pass as a serious candidate for reason, given the stage of cultural evolution we happen to be at. The evidence for the thesis that man is persistently a reason-seeker is a bit sobering; part of the evidence is that he has resorted to astrology and the like, so strong is his desire for rational-seeming authority over deliberation. Then my kind of pragmatism sees man not simply as an agent, but as a would-be rational agent: A deliberator. I am sure that what we now take for rationality in deliberation will appear rather seriously flawed, in the not-too-distant future. We are in something like the situation of the gambler who, when it was pointed out to him that the roulette game he was playing was biased against him said, 'I know, but it is the only wheel in town.'

We cannot help but act, we cannot help but deliberate; when our deliberations are coherent, we act as if we had probabilistic degrees of belief in various propositions. Robots might be programmed to have all sorts of bizarre probabilistic belief functions, and to change them in all manner of bizarre ways in response to their sensory inputs. We are not robots; but if you wish to think of us as robots or as like robots in important respects, note that we are robots of a particular narrow range of different designs. For us, some belief functions are so far out, and some modes of change are so far out as to be inaccessible. When I walk through a soaking rain I can no more believe that it is a fine, sunny day than I can fly; and in general, in any evidential situation, I have little latitude in what I am free to believe. Indeed, I seldom *choose* my beliefs – I generally simply find I *have* them, willy nilly, and generally, that is no defect. Nor is ratiocination an exception to this rule. Perhaps I am free to deliberate or not, but when I elect to deliberate I engage in an activity which, to the extent that it is successful, will pretty much force certain partial beliefs upon me, even though I may not be able to quote explicit rules that I am following.

Similarly for observation and experiment, which are ingredients in (anyway, raw materials for) deliberation. Part of the business of learning English is the matter of learning that for certain questions there are certain fairly definite ways in which one can become an authority. Take the statement, 'The sun is shining'. Part of knowing one's way about in English is knowing that by going outdoors and opening one's eyes one will have a sensibly arrived-at degree of belief in that statement which will in all probability be close to 0 or to 1. Sense-perception has its authority, too.

Man is an explaining animal as well as a deliberating animal, and the two activities are intertwined. Part of reasonableness is a readiness to survey one's own beliefs with an explanatory eye. To some extent, this is a matter of keeping track of what led us to believe what. Example: we do well to keep some track of the kinematical history of our beliefs. Perhaps I have a succession p_0, p_1, p_2, p_3 of belief functions, where each arose from its predecessor via (K). Suppose the bases of the successive changes were A, $-A$ (I got from p_0 to p_1 by changing my degree of belief in A), B, $-B$ (to get from p_1 to p_2) and finally C, $-C$ – where in each case the rigidity conditions were met. So far, so good. But I might well wonder whether I would have got p_3, starting with p_0, by a single application of (K) if I had made all three changes at once. If not, I might do well to think the whole thing over, and make some adjustments. (What adjustments?

That depends on what the propositions A, B, C are, and on what my situation is.) There is some interesting lore, here. Suppose the changes were, $p_0(A) \rightarrow p_1(A) = a$; $p_1(B) \rightarrow p_2(B) = b$; $p_2(C) \rightarrow p_3(C) = c$ (with $1-a$, $1-b$, $1-c$ as the corresponding values for the denials, of course.) If the rigidity conditions were met and A, B, C were independent relative to p_0 in the sense that $p_0(A \& B) = p_0(A)\, p_0(B)$, $p_0(A \& B \& C) = p_0(A)$ $p_0(B)\, p_0(C)$, etc., then those three must be independent relative to the other three belief functions as well, and we can get directly from p_0 to p_3 via (K), taking $n=8$, $A_1 = A \& B \& C$, $A_2 = A \& B \& -C$, ..., $A_8 = -A \& -B \& -C$, and $p_3(A \& B \& C) = abc$, $p_3(A \& B \& -C) = ab(1-c), \ldots$, $p_3(-A \& -B \& -C) = (1-a)\ (1-b)\ (1-c)$. (Of course, to apply formula (K) above, the subscript '1' should be changed to '3' throughout.)

But if A, B, C are not independent in this way, this cannot be expected to work. In that case, if the new values are a, b, c again, there is considerable latitude within which the new values of $A \& B \& -C$, etc., might lie. And by *first* changing degree of belief in A to a, *then* changing that in B to b, and only then changing that in C to c, we may well be in a position where, although we do have c as the value of $p_3(C)$, we do not have $p_3(B) = b$ (rather, $p_2(B) = b$, and the change $p_2(C) \rightarrow p_3(C) = c$ may have induced a change from $p_2(B) = b$ to $p_3(B) =$ something else) and we do not have $p_3(A) = a$. This is a reflection of the fact that since no special logical or probabilistic relationship is postulated, among A, B, and C, they do not themselves form a basis for a single jump from p_0 to p_3. The relevant basis would consist of such of the eight conjunctions of those letters and denials of those letters as have positive probability, relative to p_0. It would be idle to seek a rule telling us how to assign new probabilities to those conjuncts, given only the new probabilities of A, B, and C: Any such rule would work only in special cases, e.g. the rule that we ought to assign abc, $ab(1-c), \ldots, (1-a)\,(1-b)\,(1-c)$ respectively to $A \& B \& C$, $A \& B \& -C, \ldots, -A \& -B \& -C$ would be appropriate if and only if A, B, and C are fully independent, relative to p_0.[8]

One final comment. Part of the business of keeping an explanatory eye on our beliefs (our own and those of other agents) is, as I say, the matter of keeping track of the forces that initiate kinematical changes – changes via *K*. In one's own case, one may have a clear notion of what the basis was, in a certain case, i.e., of what the propositions $A_1, \ldots, A_n$ were, in which the change initiated. Now Levi falsely says that (aside from such private insights, I suppose),

> No shift to some nonextreme probability value can be marked off as initial by appeal to the value reached. Jeffrey lacks a way of identifying initial shifts to be used to justify other shifts ([11], p. 204).

It is the conjunction of these two statements I am calling false: the first conjunct is true enough, but Levi seems to suggest that its truth explains the truth of the second conjunct, which it cannot, since the second conjunct is false. Suppose p_0 and p_1 are successive belief functions and suppose that, in the course of examining them with an explanatory eye, I notice the existence of a partitioning $A_1, \ldots, A_n$ which meets the rigidity conditions. Then p_1 is related to p_0 as in (K) – assuming that both functions are coherent. I will then do well to explore the hypothesis that the changes $p_0(A_1) \to p_1(A_1), \ldots, p_0(A_n) \to p_1(A_n)$ were the initiating changes. The hypothesis might be false, of course; in testing it, I would want to think about the propositions $A_1, \ldots, A_n$ to see on what sensible basis the agent in question might have changed his degrees of belief in them. But perhaps I shall notice that they are all reports on the color of some disc which, at the time of the change, the agent was regarding through sunglasses. I would then be pretty confident of the explanatory hypothesis, that the A's were the basis for the change. On the other hand, I might notice that by choosing $A'_1 = A_1 \vee A_2$, $A'_2 = A_3 \vee A_4 \vee A_5$ (suppose $n=5$), I again get a basis, consisting of the simpler set, A'_1, A'_2. In such a case I would become interested in the hypothesis that the fellow was only concerned with a rather coarse color observation; that he has two categories, not five. Of course it *might* be that he had five categories in mind, and as luck would have it, the change worked out precisely as it would have done if he had had only the two in mind; but in this as in any case of explanation we are attracted by simplicity, and in the absence of evidence to the contrary might opt for the simpler explanation as likelier to be true and, at the same time, making a tighter explanatory package.

V. CONCLUSION

Wolfman and Dracula live in disjoint, unreal worlds; when Dracula meets Wolfman in the movies, the two fantasies generally destroy each other –

Wolfman cannot breathe the musty air of Castle Dracula, nor can Dracula survive the winds of Wolfman's moors. Now that Levi and I have confronted each other, I see that my title may have been inappropriate: perhaps our worlds can mix, without reducing each other to absurdity. Perhaps, indeed, the disparity between them is the disparity of viable accounts of different parts of the real world, not that of incompatible fantasies.

Before the confrontation my dissatisfaction with Levi's positive proposals – his Bayesian account [10] of the business of accepting and rejecting hypotheses – was centered on the very notions of acceptance and rejection. I observed [9] that while he gave methods for deciding which (if either) of the two acts *accept H* and *reject H* one ought to perform, he provided no account of how one is to go about performing those acts. The notions of belief and disbelief are familiar enough but, I find, unclear. In contrast, I find the notion of subjective probability, for all its (decreasing) unfamiliarity, to be a model of clarity – a clarity that it derives from its association with the concepts of utility and preference within the framework of Bayesian decision theory. But as I understand him, Levi now takes the notions of acceptance and rejection to be in need of clarification, which he seeks to provide by giving them a place in an account of inquiry of the sort discussed in his contribution to this conference. I take it that this account is still in the programmatic stage, and I await its elaboration with interest. It would be good to make systematic sense of our talk of acceptance and rejection of hypotheses, and the *prima facie* difficulty of giving a coherent account is no reason for thinking that no such account will be forthcoming. Meanwhile, in trying to square theory with practice, *I* continue to avoid talk about knowledge and acceptance of hypotheses, trying to make do with graded belief, as in [8]. That, too, is programmatic, but it is the program that strikes me as most promising. For both programs I think the crucial task is that of squaring a Bayesian account of practical deliberation, which both Levi and I seem to accept, with the facts about how theory is generally thought to legitimately impinge upon practice. 'My kind of pragmatism' would be shattered by a wedge driven between theory and practice.[9]

University of Pennsylvania

BIBLIOGRAPHY

[1] G. E. Bates, *Probability*, Reading, Mass., 1965.

[2] Ethan Bolker, *Functions Resembling Quotients of Measures*, Dissertation, Harvard University, 1965.

[3] Ethan Bolker, 'Functions Resembling Quotients of Measures', *Trans. American Mathematical Society* **124** (1966) 292–312.

[4] Ethan Bolker, 'A Simultaneous Axiomatization of Utility and Subjective Probability', *Philosophy of Science* **34** (1967) 333–40.

[5] M. Friedman and L. J. Savage, 'The Utility Analysis of Choices Involving Risk', *Journal of Political Economy* **56** (1948) 279–304.

[6] Ian Hacking, 'Slightly More Realistic Personal Probability', *Philosophy of Science* **34** (1967) 311–25.

[7] Richard C. Jeffrey, *The Logic of Decision*, New York 1965.

[8] Richard C. Jeffrey, 'Probable Knowledge', in *The Problem of Inductive Logic* (ed. by I. Lakatos), Amsterdam 1968, pp. 166–80.

[9] Richard C. Jeffrey, 'Review of I. Levi's *Gambling With Truth*', *Journal of Philosophy* **65** (1968) 313–22.

[10] Isaac Levi, *Gambling With Truth*, New York and London 1967.

[11] Isaac Levi, 'Probability Kinematics', *British Journal for the Philosophy of Science* **18** (1967) 197–209.

[12] Howard Riaffa, *Decision Analysis: Introductory Lectures on Choices under Uncertainty*, Reading, Mass., 1968.

[13] F. P. Ramsey, *The Foundations of Mathematics and other Logical Essays*, London and New York 1931.

[14] Leonard J. Savage, *The Foundations of Statistics*, New York 1954.

[15] Patrick Suppes, 'Rational Changes of Belief', in *The Problem of Inductive Logic* (ed. by I. Lakatos), Amsterdam 1968, pp. 186–8.

[16] Richard Todd, 'Where Money Grows on Decision Trees', *New York Times Magazine*, November 10, 1968.

[17] I. Todhunter, *A History of the Mathematical Theory of Probability*, New York 1949.

REFERENCES

1 Those issues were indeed greatly clarified during the conference; but it seems best to publish the paper just as it was presented at the conference even unto the title – except for minor corrections throughout, and for some brief concluding remarks in Section V.

2 'My' account is in [7]; it is a modification and, as I see it, an improvement of the accounts of Ramsey [13] and Savage [14].

3 See [2], [3], and [4]. Bolker's work was prior to mine: A case in which pure mathematics had an unexpected application.

4 For some striking illustrations of the flexibility of the Bayesian framework, see [5], where some puzzling and rather complex behavior is rationalized via a simple hypothesis about the shape of the utility curve for income.

5 This is the celebrated argument from coherence: If his degrees of belief do not satisfy the probability axioms there will be a set of bets, each of which looks fair to the agent, but on which he will with logical necessity suffer an overall loss. See [7], p. 49 and references given there.

[6] The room in which this was read had some 40 people in it. Imagine that the date was September 25.
[7] *Politics* 1260^{a} 13.
[8] One may take this as a reply to the objection, 'One difficulty immediately leaps to the eye...', just below the formula in [11], p. 205.
[9] In [10], p. 13, Levi observes that his "critical cognitivism renders asunder, at least partially, what many philosophers have endeavored to join together – theoretical and practical wisdom".

RALPH L. SLAGHT

INDUCTION, ACCEPTANCE, AND RATIONAL BELIEF: A SELECTED BIBLIOGRAPHY

This bibliography is constructed, in the main, of books and articles published in the last 20 years, which pertain to or are related to the topic of the book. The bibliography has four sections: (I) Books and Anthologies, (II) Selected Reviews of those books, (III.A) Articles pertaining to the topic of the book and (III.B) Articles on related topics. No section contains an exhaustive compilation of materials, although an attempt was made not to omit any major contribution in the first three sections. Section III.B deals chiefly, but not exclusively, with articles on paradoxes of confirmation, including Goodman's new riddle of induction and Miller's paradox of information; the problems of confirmation theory and attempts to define measures of degrees of confirmation; simplicity; explanation; the logic of knowledge and belief; and analyses of empirical knowledge. For bibliographies of earlier material, see Keynes [50], Carnap [16], and von Wright [118]; for more recent literature see Kyburg [55] and [571] and Hilpinen [356].

I. BOOKS AND ANTHOLOGIES

[1] R. Ackermann, *Nondeductive Inference*, New York 1966.

[2] G. E. M. Anscombe, *Intention*, 2nd edition, Ithaca 1963. (For reviews, see Jarvis [197], Copi [162], and Chisholm [153].)

[3] A. J. Ayer, *Foundations of Empirical Knowledge*, New York 1940. (For reviews, see Buchler [144] and Church [155].)

[4] A. J. Ayer, *Philosophical Essays*, London 1954. (For reviews, see Barnes [130], Baylis [133], Chisholm [152], and Cohen [159].)

[5] A. J. Ayer, *The Problem of Knowledge*, New York 1955. (For a review, see White [306].)

[6] A. J. Ayer, *Language, Truth, and Logic*, 2nd edition, New York 1952. (For reviews, see Nagel [253], Church [156].)

[7] Y. Bar-Hillel (ed.), *Proceedings of the 1964 Congress for Logic, Methodology, and the Philosophy of Science*, Amsterdam 1965.

[8] Stephen Barker, *Induction and Hypothesis: A Study of the Logic of Confirmation*, Ithaca 1957. (For reviews, see Salmon [275], Rosthal [271], Hirst [195], Harré [179], and Svenonius [297].)

[9] G. E. Bates, *Probability*, Reading, Mass., 1965.

[10] B. Baumrin (ed.), *Philosophy of Science: The Delaware Seminar*, vol. II, New York 1963. (For a review, see Hirschmann [194].)

[11] Emile Borel (ed.), *Traité du calcul des probabilités et de ses applications*, 4 vols., Paris 1925ff.
[12] R. B. Braithwaite, *Scientific Explanation*, Cambridge 1953. (For reviews, see Woodger [312], Nagel [261], Jeffreys [202], and Shimony [284].)
[13] C. D. Broad, *Induction, Probability, and Causation*, Dordrecht 1968.
[14] R. Carnap, *The Nature and Application of Inductive Logic*, Chicago 1951.
[15] R. Carnap, *The Continuum of Inductive Methods*, Chicago 1952. (For reviews, see Hay [181], Burks [146], and Russell [273].)
[16] R. Carnap, *Logical Foundations of Probability*, 2nd edition, Chicago 1962. (For reviews, see Kyburg [216], Kemeny [207] and [208], Ceccato [151], Cousin [164], Dantzig [165], Toulmin [298], Vincent [302], and Burks [145].)
[17] R. Carnap and Wolfgang Stegmüller, *Induktive Logik und Wahrscheinlichkeit*, Wien 1959. (For a review, see Bar-Hillel [125].)
[18] R. Chisholm, *Perceiving: A Philosophical Study*, Ithaca 1959. (For reviews, see Baylis [134] and [135], Yolton [313], and Price [265].)
[19] R. Chisholm, *Theory of Knowledge*, Englewood Cliffs, N.J., 1966. (For reviews, see O'Hair [263], Joske [203], and McQueen [236].)
[20] R. Chisholm *et al.*, *Philosophy*, Englewood Cliffs, N.J., 1964. (For reviews, see Melchert [245], Nakhnikian [262], and Quinton [266].)
[21] C. West Churchman, *Theory of Experimental Inference*, New York 1948. (For reviews, see Hempel [184], and Spilsbury [288].)
[22] R. Chisholm, *Prediction and Optimal Decision*, Englewood Cliffs, N.J., 1962. (For reviews, see Beers [137] and Kyburg [215].)
[23] R. S. Cohen and M. W. Wartofsky (eds.), *Boston Studies in the Philosophy of Science*, vol. II, New York 1965.
[24] Harald Cramér, *Mathematical Methods of Statistics*, Princeton 1945.
[25] A. Danto, *Analytical Philosophy of Knowledge*, Cambridge 1968.
[26] J. P. Day, *Inductive Probability*, New York 1961. (For reviews, see Salmon [277], Hutten [196], Fisk [171], and Tucker [299].)
[27] W. Edwards and A. Tversky (eds.), *Decision Making*, Baltimore 1967.
[28] H. Feigl and G. Maxwell, *Minnesota Studies in the Philosophy of Science*, vol III, Minneapolis 1962. (For reviews, see Achinstein [121], Hesse [189], Madden [239], and Smart [287].)

[29] H. Feigl and G. Maxwell (eds.) *Current Issues in the Philosophy of Science*, New York 1961. (For reviews, see Hesse [188] and Smart [286].)

[30] Paul K. Feyerabend and G. Maxwell (eds.), *Mind, Matter, and Method: Essays in Philosophy and Science in Honor of Herbert Feigl*, Minneapolis 1966 (especially Part II: 'Induction, Confirmation, and Philosophic Method'). (For reviews, see Louch [233] and Young *et al.* [314].)

[31] Ronald A. Fisher, *Statistical Methods and Scientific Inference*, Edinburgh 1959. First edition, 1956.

[32] M. H. Foster and M. L. Martin (eds.), *Probability, Confirmation and Simplicity*, New York 1966.

[33] P. G. Frank (ed.), *The Validation of Scientific Theories*, Boston 1954.

[34] I. J. Good, *Probability and the Weighing of Evidence*, London 1950.

[35] I. J. Good, *The Estimation of Probabilities: An Essay on Modern Bayesian Methods*, Boston 1965. (For a review, see Levi [227].)

[36] N. Goodman, *Fact, Fiction and Forecast*, Cambridge, Mass., 1955, 2nd edition, New York 1965. (For reviews, see Cooley [160], Madden [237], Martin [242], Pap [264], Stove [293], and Watling [304].)

[37] A. P. Griffiths (ed.), *Knowledge and Belief*, London 1967.

[38] Ian Hacking, *Logic of Statistical Inference*, Cambridge 1965. (For reviews, see Levi [227], Clendinnen [157], and Moss [250].)

[39] R. F. Harrod, *The Foundations of Inductive Logic*, New York 1956. (For reviews, see Levi [223], Stove [294], and Hamlyn [176].)

[40] C. G. Hempel, *Aspects of Scientific Explanation*, New York 1965. (For reviews, see Hesse [193] and Resnik [268].)

[41] Jaakko Hintikka, *Knowledge and Belief*, Ithaca 1962. (For reviews, see Chisholm [154], Geach [172], White [305], Castañeda [147], Lemmon [222], and Deutcher [167].)

[42] Jaakko Hintikka and P. Suppes (eds.), *Aspects of Inductive Logic*, Amsterdam 1966. (For reviews, see Levi [288] and Fenstad [170].)

[43] Jaakko Hintikka and P. Suppes (eds.), *Information and Inference*, Dordrecht 1969 (forthcoming).

[44] David Hume, *A Treatise on Human Nature, Being an Attempt to Introduce the Experimental Method of Reasoning into Moral Subjects*, London 1739.

[45] David Hume, *An Enquiry Concerning Human Understanding*, London 1748.

[46] R. C. Jeffrey, *The Logic of Decision*, New York 1965. (For reviews, see Levi [227], Robinson [270], and Schick [280].)

[47] H. Jeffreys, *Theory of Probability*, London 1939. (For reviews, see Nagel [251] and Koopman [212].)

[48] H. Jeffreys, *Scientific Inference*, 2nd edition, London 1957. First edition, 1931. (For reviews, see Koslow [214], Hesse [186], Barker [127], Robinson [269], Nagel [255], and Salmon [274].)

[49] Jerrold J. Katz, *The Problem of Induction and Its Solution*, Chicago 1962. (For reviews, see Schick [279], Barker [128], Baumer [136], Hacking [173], Harré [180], Hesse [191], Stoothoff [292], and Stove [295].)

[50] John M. Keynes, *A Treatise on Probability*, London 1921.

[51] R. Klibansky (ed.). *Philosophy in Mid-Century*. Part I: *Logic*, Firenze 1958.

[52] William Kneale, *Probability and Induction*, London 1949. (For reviews, see Broad [143], Will [309], and Nagel [260].)

[53] A. N. Kolmogoroff, *Foundations of the Theory of Probability*, New York 1950.

[54] Stephen Körner (ed.), *The Colston Papers 9, Observation and Interpretation: A Symposium of Philosophers and Scientists*, New York 1957. (For a review, see Scriven [281].)

[55] H. E. Kyburg, Jr., *Probability and the Logic of Rational Belief*, Middletown 1961. (For a review, see Salmon [278].)

[56] H. E. Kyburg, Jr. and E. Nagel (eds.), *Induction: Some Current Issues*, Middletown, Conn., 1963. (For reviews, see Hacking [174] and Caws [150].)

[57] H. E. Kyburg, Jr. and Howard Smokler (eds.) *Readings in Subjective Probability*, New York 1964.

[58] I. Lakatos (ed.), *The Problem of Inductive Logic*, Amsterdam 1968. (For a review, see Levi [757].)

[59] I. Lakatos and A. Musgrave (eds.), *Problems in the Philosophy of Science*, Amsterdam 1968. (For a review, see N. Maxwell [243].)

[60] Pierre Simon de Laplace, *Théorie analytique des probabilités*, Paris 1812. 2nd edition 1814. 3rd edition 1820. (Reprinted in: *Oeuvres Complètes*, vol. VII, Paris 1847.)

[61] Pierre Simon de Laplace, *Essai philosophique sur les probabilités*, in

Oeuvres Complètes, vol. VII, edition in 2 vols. (Reprinted from 5th edition 1825), Paris 1921. (German transl. by R. V. Misis, Leipzig 1932; English transl. by Truscott and Emory, New York 1902.)

[62] H. Leblanc, *Statistical and Inductive Probabilities*, Englewood Cliffs, N.J., 1962. (For reviews, see Kyburg [217], Levi [226], and Michal [247].)

[63] Daniel Lerner (ed.), *Evidence and Inference*, Glencoe 1959. (For a review, see Abelson [120].)

[64] Isaac Levi, *Gambling With Truth: An Essay on Induction and the Aims of Science*, New York 1967. (For reviews, see Jeffrey [199], Hacking [175], Mackie [235], and Lehrer [221].)

[65] C. I. Lewis, *Mind and the World Order*, New York 1956 (reprint of 1929). (For a review, see Baylis [131].)

[66] C. I. Lewis, *An Analysis of Knowledge and Valuation*, La Salle 1946. (For reviews, see Henle [185], Hempel [183], Stace [289], Baylis [132], Ducasse [168], and Long [232].)

[67] R. D. Luce and H. Raiffa, *Games and Decisions*, New York 1957.

[68] D. V. Lindley, *Introduction to Probability and Statistics from a Bayesian Viewpoint*, vol. II: *Inference*, Cambridge 1965.

[69] E. H. Madden, *The Structure of Scientific Thought*, Boston 1960. (For reviews, see Levi [225] and Tucker [300].)

[70] Norman Malcolm, *Knowledge and Certainty*, Englewood Cliffs, N.J., 1963. (For reviews, see Castañeda [148], Hamlyn [177], and Katsoff [204].)

[71] M. L. Martin, (see Foster [32].)

[72] R. M. Martin, *Truth and Denotation: A Study in Semantical Theory*, Chicago 1958.

[73] R. M. Martin, *Toward a Systematic Pragmatics*, Amsterdam 1959.

[74] R. M. Martin, *Intention and Decision*, Englewood Cliffs, N.J., 1963. (For reviews, see Keene [206], Kyburg [218], Levison [231], and Vickers [301].)

[75] G. Maxwell, (see Feyerabend [30].)

[76] G. Maxwell, (see Feigl [28] and [29]).

[77] John Stuart Mill, *A System of Logic, Ratiocinative and Inductive, Being a Connected View of the Principles of Evidence and the Methods of Scientific Investigation*, 2 vols., London 1843. 8th edition New York 1930.

[78] Richard von Mises, *Probability, Statistics and Truth*, New York 1957 (transl. of German edition, revised 1951). (For reviews, see Nagel [254] and Stannard [290].)
[79] S. Morgenbesser (ed.), *Philosophy of Science Today*, New York 1967.
[80] S. Morgenbesser and A. Danto (eds.), *Philosophy of Science*, New York 1960. (For a review, see Levi [224].)
[81] E. Nagel, *Principles of the Theory of Probability*, Chicago 1939. (For reviews, see Costello [163], Langford [220], and von Mises [248].)
[82] E. Nagel, *The Structure of Science*, New York 1961. (For reviews, see Caton [149], Harré [178], Hesse [187], Keene [205], and Scriven [282].)
[83] Nagel *et al.* (eds.), *Logic, Methodology and Philosophy of Science*, Stanford 1962. (For reviews, see Black [141], Hesse [190], and Mendelson *et al.* [246].)
[84] E. Nagel, (see Kyburg [56]).
[85] Emanuel Parzen, *Modern Probability Theory and its Applications*, New York 1960.
[86] C. S. Peirce, *Collected Papers*, vols. I–VI (ed. by Hartshorne and Weiss), Cambridge 1932–35. Vols. VII–VIII (ed. by Burks), Cambridge 1958. (For a review of vols. I–VI, see Nagel [252].)
[87] K. R. Popper, *The Logic of Scientific Discovery*, London 1959 (first published as *Logik der Forschung* in 1934). (For reviews, see Fain [169], Jeffrey [198], Rescher [267], and Warnock [303].)
[88] K. R. Popper, *Conjectures and Refutations*, London 1963. (For reviews, see Alexander [123], Ayer [124], Levison [230], Achinstein [122], and Hesse [192].)
[89] H. A. Prichard, *Knowledge and Perception*, Oxford 1950.
[90] H. Raiffa, *Decision Analysis: Introductory Lectures on Choices under Uncertainty*, Reading, Mass., 1968.
[91] H. Raiffa and R. Schlaifer, *Applied Statistical Decision Theory*, Cambridge, Mass., 1961.
[92] F. P. Ramsey, *The Foundations of Mathematics and Other Logical Essays*, London and New York 1931.
[93] Hans Reichenbach, *The Theory of Probability*, Berkeley and Los Angeles 1949. (For reviews, see Williams [311], Jeffreys [200], and Nagel [259].)
[94] Bertrand Russell, *Human Knowledge: Its Scope and Limits*, New

York 1948. (For reviews, see Bergmann [138], Kneale [210], Malcolm [241], Maziarz [244], and L. J. Russell [272].)
[95] W. C. Salmon, *The Foundations of Scientific Inference*, Pittsburgh 1966–67. (For a review, see Levi [229].)
[96] Leonard J. Savage, *Foundations of Statistics*, New York 1954. (For a review, see Shapiro [283].)
[97] Leonard J. Savage *et al.*, *The Foundations of Statistical Inference*, New York 1962.
[98] I. Scheffler, *Anatomy of Inquiry*, New York 1963. (For reviews, see Barker [129], Cohen [158], Mackie [234], Madden [240], Stenner [291], and Stove [296].)
[99] I. Scheffler, *Conditions of Knowledge: An Introduction to Epistemology and Education*, Chicago 1966. (For a review, see O'Hair [263].)
[100] Paul A. Schilpp (ed.), *The Philosophy of Rudolph Carnap* (The Library of Living Philosophers, vol. XI), La Salle, Ill., 1968. (For reviews, see Kyburg [219] and Moss [249].)
[101] G. L. S. Shackle, *Decision, Order and Time*, Cambridge 1961.
[102] B. Skyrms, *Choice and Chance*, Belmont 1966. (For a review, see Silver [285].)
[103] H. Smokler, (see Kyburg [57]).
[104] W. Stegmüller, (see Carnap [17]).
[105] P. Suppes, (see Hintikka [42]).
[106] P. Suppes, (see Nagel [83]).
[107] A. Tarski, (see Nagel [83]).
[108] Håkan Törnebohm, *Information and Confirmation* (Gothenburg Studies in Philosophy, No. 3), Goteberg 1964. (For a review, see Levi [227].)
[109] S. Toulmin, *The Uses of Argument*, Cambridge 1958. (For reviews, see Cooley [161], Will [310], and Körner [213].)
[110] B. Van Rootselaar and J. Staal (eds.), *Proceedings of the 3rd International Congress for Logic, Methodology and Philosophy of Science*, Amsterdam 1968.
[111] John Venn, *The Logic of Chance. An Essay on the Foundations and Province of the Theory of Probability, with Especial Reference to its Logical Bearings and its Application to Moral and Social Science and to Statistics*, London and New York 1866, and later editions. (For a review, see Kneale [211].)

[112] John Von Neumann and Oskar Morgenstern, *Theory of Games and Economic Behavior*, Princeton 1947. (For a review, see Nagel [257].)

[113] Abraham Wald, *Statistical Decision Functions*, New York 1960.

[114] M. White (ed.), *Academic Freedom, Logic, and Religion*, Philadelphia 1953.

[115] Donald C. Williams, *The Ground of Induction*, Cambridge 1947. (For reviews, see Nagel [258], Will [308], and Black [140].)

[116] John Culton Wisdom, *Foundations of Inference in Natural Science*, London 1952. (For reviews, see Whitely [307] and Beveridge [139].)

[117] G. H. von Wright, *A Treatise on Induction and Probability*, New York 1951. (For reviews, see Jeffreys [201], Hay [182], and Kemeny [209].)

[118] G. H. von Wright, *The Logical Problem of Induction*, 2nd revised edition, New York 1957. (For reviews, see Day [166], Nagel [256], Broad [142], Barker [126], Salmon [276], and Madden [238].)

[119] W. Yourgrau (ed.), *Logic, Physical Reality and History*, New York (forthcoming).

II. REVIEWS

[120] R. Abelson, Review of Lerner [63], *Philosophy and Phenomenological Research* **21** (1960–61) 413–4.

[121] P. Achinstein, Review of Feigl and Maxwell [28], *Philosophical Review* **73** (1964) 106.

[122] P. Achinstein, Review of Popper [88], *The British Journal for the Philosophy of Science* **19** (1968) 159.

[123] P. Alexander, Review of Popper [88], *Philosophical Books* **4** (1963) 23.

[124] A. J. Ayer, Review of Popper [88], *New Statesman* **65** (1963) 155.

[125] Y. Bar-Hillel, Review of Carnap and Stegmüller [17], *Philosophy of Science* **29** (1962) 94–5.

[126] S. F. Barker, Review of von Wright [118], *Journal of Philosophy* **55** (1958) 130–1.

[127] S. F. Barker, Review of Jeffreys [48], *Philosophical Review* **67** (1958) 404–7.

[128] S. F. Barker, Review of Katz [49], *Philosophical Review* **73** (1964) 111.

[129] S. F. Barker, Review of Scheffler [98], *Journal of Philosophy* **63** (1966) 358.
[130] W. H. F. Barnes, Review of Ayer [4], *Philosophy* **32** (1957) 67–70.
[131] C. A. Baylis, Review of Lewis [65], *Journal of Philosophy* **27** (1930) 320–7.
[132] C. A. Baylis, Review of Lewis [66], *Philosophy and Phenomenological Research* **8** (1947) 152–9.
[133] C. A. Baylis, Review of Ayer [4], *Philosophical Review* **64** (1955) 640–4.
[134] C. A. Baylis, Review of Chisholm [18], *Journal of Philosophy* **56** (1959) 773–91.
[135] C. A. Baylis, Review of Chisholm [18], *Journal of Symbolic Logic* **27** (1962) 123–4.
[136] W. H. Baumer, Review of Katz [49], *Philosophy and Phenomenological Research* **25** (1964–65) 295.
[137] S. Beers, Review of Churchman [22], *Philosophy of Science* **30** (1963) 84–9.
[138] G. Bergmann, Review of Russell [94], *Physics Today* **2** (1949) 27–8.
[139] W. I. B. Beveridge, Review of Wisdom [116], *The British Journal for the Philosophy of Science* **3** (1952–53) 291–3.
[140] M. Black, Review of Williams [115], *Journal of Symbolic Logic* **12** (1947) 141–4.
[141] M. Black, Review of Nagel *et al.* [83], *Philosophical Review* **72** (1963) 538.
[142] C. D. Broad, Review of von Wright [118], *Mind* **53** (1944) 1–24, 97–119, and 193–214.
[143] C. D. Broad, Review of Kneale [52], *Mind* **59** (1950) 94–115.
[144] J. Buchler, Review of Ayer [3], *Journal of Philosophy* **38** (1941) 219–20.
[145] Burks, Review of Carnap [16], *Journal of Philosophy* **48** (1951) 524–35.
[146] Burks, Review of Carnap [15], *Journal of Philosophy* **50** (1953) 731–4.
[147] H. N. Castañeda, Review of Hintikka [41], *Journal of Symbolic Logic* **29** (1964) 132.
[148] H. N. Castañeda, Review of Malcolm [70], *Review of Metaphysics* **18** (1964–65) 508–47.

[149] C. E. Caton, Review of Nagel [82], *Philosophical Review* **73** (1964) 104.
[150] P. Caws, Review of Kyburg [56], *Philosophy and Phenomenological Research* **25** (1964–65) 427–8.
[151] S. Ceccato, Review of Carnap [16], *Methodos* **6** (1954) 153–4.
[152] R. M. Chisholm, Review of Ayer [4], *Journal of Philosophy* **52** (1955) 189–95.
[153] R. M. Chisholm, Review of Anscombe [2], *Philosophical Review* **68** (1959) 110–15.
[154] R. M. Chisholm, Review of Hintikka [41], *Journal of Philosophy* **60** (1963) 773–95.
[155] A. Church, Review of Ayer [3], *Journal of Symbolic Logic* **6** (1941) 108.
[156] A. Church, Review of Ayer [6], *Journal of Symbolic Logic* **14** (1949) 52–3.
[157] J. Clendinnen, Review of Hacking [38], *Australasian Journal of Philosophy* **44** (1966) 264.
[158] L. J. Cohen, Review of Scheffler [98], *Philosophical Quarterly* **15** (1965) 374.
[159] M. Cohen, Review of Ayer [4], *Mind* **65** (1956) 563–5.
[160] J. C. Cooley, Review of Goodman [36], *Journal of Philosophy* **54** (1957) 293–311.
[161] J. C. Cooley, Review of Toulmin [109], *Journal of Philosophy* **56** (1959) 297.
[162] I. M. Copi, Review of Anscombe [2], *Philosophy of Science* **26** (1959) 148–9.
[163] H. T. Costello, Review of Nagel [81], *Journal of Philosophy* **36** (1939) 163–5.
[164] D. R. Cousin, Review of Carnap [16], *Philosophical Quarterly* **4** (1954) 82–4.
[165] D. van Dantzig, Review of Carnap [16], *Synthese* **8** (1953) 459–70.
[166] J. P. Day, Review of von Wright [118], *Philosophy* **35** (1960) 77–80.
[167] M. Deutscher, Review of Hintikka [41], *Mind* **75** (1966) 145–9.
[168] C. J. Ducasse, Review of Lewis [66], *Philosophical Review* **57** (1948) 260–80.
[169] H. Fain, Review of Popper [87], *Philosophy of Science* **28** (1961) 319–24.

[170] J. E. Fenstad, Review of Hintikka and Suppes [42], *Synthese* **17** (1967) 449–60.

[171] Fisk, Review of Day [26], *Philosophical Studies* (Dublin) **11** (1961–62).

[172] P. T. Geach, Review of Hintikka [41], *Philosophical Books* **4** (1963) 7.

[173] I. Hacking, Review of Katz [49], *The British Journal for the Philosophy of Science*, **15** (1964–65) 345.

[174] I. Hacking, Review of Kyburg [56], *The British Journal for the Philosophy of Science*, **15** (1964–65) 347–8.

[175] I. Hacking, Review of Levi [64], *Synthese* **17** (1967) 444–8.

[176] D. W. Hamlyn, Review of Harrod [39], *Philosophy* **33** (1958) 369–70.

[177] D. W. Hamlyn, Review of Malcolm [70], *Philosophy* **40** (1965) 169.

[178] R. Harré, Review of Nagel [82], *Philosophical Books* **2** (1961) 19.

[179] R. Harré, Review of Barker [8], *Mind* **71** (1962) 412–20.

[180] R. Harré, Review of Katz [49], *Mind* **73** (1964) 457.

[181] W. H. Hay, Review of Carnap [15], *Philosophical Review* **62** (1953) 468–72.

[182] W. H. Hay, Review of von Wright [117], *Journal of Philosophy* **50** (1953) 782–8.

[183] C. G. Hempel, Review of Lewis [66], *Journal of Symbolic Logic* **13** (1948) 40–5.

[184] C. G. Hempel, Review of Churchman [21], *Journal of Philosophy* **46** (1949) 557–61.

[185] P. Henle, Review of Lewis [66], *Journal of Philosophy* **45** (1948) 524–32.

[186] M. Hesse, Review of Jeffreys [48], *Philosophy* **34** (1959) 66–8.

[187] M. Hesse, Review of Nagel [82], *Mind* **72** (1963) 429.

[188] M. Hesse, Review of Feigl and Maxwell [29], *Review of Metaphysics* **17** (1963–64) 98–108.

[189] M. Hesse, Review of Feigl and Maxwell [28], *Review of Metaphysics* **17** (1963–64) 98–108.

[190] M. Hesse, Review of Nagel *et al.* [83], *Review of Metaphysics* **18** (1964–65) 109–22.

[191] M. Hesse, Review of Katz [49], *Review of Metaphysics* **18** (1964–65) 109–22.

[192] M. Hesse, Review of Popper [88], *Philosophical Quarterly* **15** (1965) 372.

[193] M. Hesse, Review of Hempel [40], *The British Journal for the Philosophy of Science* **18** (1967) 67.

[194] D. Hirschmann, Review of Baumrin [10], *The British Journal for the Philosophy of Science* **17** (1966) 155.

[195] P. H. Hirst, Review of Barker [8], *Philosophical Quarterly* **10** (1960) 375–6.

[196] E. H. Hutten, Review of Day [26], *Mind* **71** (1962) 583.

[197] J. Jarvis, Review of Anscombe [2], *Journal of Philosophy* **56** (1959) 31–41.

[198] R. C. Jeffrey, Review of Popper [87], *Econometrica* **28** (1960) 925.

[199] R. C. Jeffrey, Review of Levi [64], *Journal of Philosophy* **65** (1968) 313–22.

[200] H. Jeffreys, Review of Reichenbach [93], *Nature* **166** (1950) 413–4.

[201] H. Jeffreys, Review of von Wright [117], *The British Journal for the Philosophy of Science* **3** (1952–53) 276–7.

[202] H. Jeffreys, Review of Braithwaite [12], *The British Journal for the Philosophy of Science* **4** (1953–54) 348–9.

[203] W. D. Joske, Review of Chisholm [19], *Australasian Journal of Philosophy* **44** (1966) 394.

[204] L. D. Kattsoff, Review of Malcolm [70], *Philosophy and Phenomenological Research* **26** (1965–66) 263.

[205] G. B. Keene, Review of Nagel [82], *Philosophy* **37** (1962) 372.

[206] G. B. Keene, Review of Martin [74], *Philosophical Quarterly* **16** (1966) 83.

[207] J. G. Kemeny, Review of Carnap [16], *Journal of Symbolic Logic* **16** (1951) 205–7.

[208] J. G. Kemeny, Review of Carnap [16], *Review of Metaphysics* **5** (1951–52) 145–56.

[209] J. G. Kemeny, Review of von Wright [117], *Philosophical Review* **62** (1953) 93–101.

[210] W. Kneale, Review of Russell [94], *Mind* **58** (1949) 369–87.

[211] W. Kneale, Review of Venn [111], *The British Journal for the Philosophy of Science* **14** (1963–64) 73–4.

[212] B. O. Koopman, Review of Jeffreys [47], *Journal of Symbolic Logic* **8** (1943) 34.

[213] S. Körner, Review of Toulmin [109], *Mind* **68** (1959) 425.
[214] A. Koslow, Review of Jeffreys [48], *Journal of Philosophy* **57** (1960) 384–91.
[215] H. E. Kyburg, Jr., Review of Churchman [22], *Journal of Philosophy* **59** (1962) 549–54.
[216] H. E. Kyburg, Jr., Review of Carnap [16], *Journal of Philosophy* **60** (1963) 362–4.
[217] H. E. Kyburg, Jr., Review of Leblanc [62], *Philosophical Review* **73** (1964) 269.
[218] H. E. Kyburg, Jr., Review of Martin [74], *Philosophical Review* **74** (1965) 386.
[219] H. E. Kyburg, Jr., Review of Schilpp [100], *Journal of Philosophy* **65** (1968) 503–15.
[220] C. H. Langford, Review of Nagel [81], *Journal of Symbolic Logic* **4** (1939) 119–20.
[221] K. Lehrer, Review of Levi [64], *Nous* **3** (1969) 255–97.
[222] E. J. Lemmon, Review of Hintikka [41], *Philosophical Review* **74** (1965) 381.
[223] I. Levi, Review of Harrod [39], *Journal of Philosophy* **55** (1958) 209–12.
[224] I. Levi, Review of Morgenbesser and Danto [80], *Journal of Philosophy* **58** (1961) 387–90.
[225] I. Levi, Review of Madden [69], *Journal of Philosophy* **58** (1961) 387–90.
[226] I. Levi, Review of Leblanc [62], *Journal of Philosophy* **60** (1963) 21–5.
[227] I. Levi, Reviews of Good [35], Hacking [38], Jeffrey [46] and Törnebohm [108], *Synthese* **16** (1966) 234–44.
[228] I. Levi, Review of Hintikka and Suppes [42], *The British Journal for the Philosophy of Science* **18** (1968) 73–81.
[229] I. Levi, Review of Salmon [95], *The British Journal for the Philosophy of Science* **19** (1968) 259.
[230] A. B. Levison, Review of Popper [88], *Science* **140** (1963) 643.
[231] A. B. Levison, Review of Martin [74], *Philosophy and Phenomenological Research* **25** (1964–65) 294.
[232] W. Long, Review of Lewis [66], *The Personalist* **30** (1949) 285–7.
[233] A. R. Louch, Review of Feyerabend and Maxwell [30], *Journal of the History of Philosophy* **5** (1967) 193.

[234] J. L. Mackie, Review of Scheffler [98], *The British Journal for the Philosophy of Science* **16** (1965–66) 57–61.

[235] J. L. Mackie, Review of Levi [64], *The British Journal for the Philosophy of Science* **19** (1968) 261.

[236] D. McQueen, Review of Chisholm [19], *Philosophical Books* **8** (1967) 4.

[237] E. H. Madden, Review of Goodman [36], *Philosophy and Phenomenological Research* **16** (1955) 271–3.

[238] E. H. Madden, Review of von Wright [118], *Philosophy and Phenomenological Research* **18** (1957–58) 550–1.

[239] E. H. Madden, Review of Feigl and Maxwell [28], *Philosophy and Phenomenological Research* **24** (1963–64) 287.

[240] E. H. Madden, Review of Scheffler [98], *Philosophy of Science* **33** (1966) 82–4.

[241] N. Malcolm, Review of Russell [94], *Philosophical Review* **59** (1950) 94–106.

[242] R. M. Martin, Review of Goodman [36], *Journal of Symbolic Logic* **25** (1960) 250–1.

[243] N. Maxwell, Review of Lakatos and Musgrave [59], *The British Journal for the Philosophy of Science* **20** (1969) 81–3.

[244] E. A. Maziarz, Review of Russell [94], *New Scholasticism* **23** (1949) 318–25.

[245] N. Melchert, Review of Chisholm *et al.* [20], *Philosophy and Phenomenological Research* **26** (1965–66) 426.

[246] E. Mendelson *et al.*, Review of Nagel *et al.* [83], *Journal of Philosophy* **61** (1964) 76–94.

[247] A. C. Michal, Review of Leblanc [62], *Philosophy of Science* **34** (1967) 195–6.

[248] R. von Mises, Review of Nagel [81], *Erkentniss* **8** (1939) 261–3.

[249] J. M. B. Moss, Review of Schilpp [100], *Philosophical Books* **6** (1965) 25–8.

[250] J. M. B. Moss, Review of Hacking [38], *Philosophical Books* **7** (1966) 12.

[251] E. Nagel, Review of Jeffreys [47], *Journal of Philosophy* **29** (1932) 409–12.

[252] E. Nagel, Reviews of Peirce [86], *Journal of Philosophy* **30** (1933) 365–86; **31** (1934) 188–90; **33** (1936) 107–9.

[253] E. Nagel, Review of Ayer [6], *Journal of Philosophy* **33** (1936) 328–30.

[254] E. Nagel, Review of von Mises [78], *Journal of Philosophy* **34** (1937) 80.

[255] E. Nagel, Review of Jeffreys [48], *Journal of Philosophy* **37** (1940) 524–8.

[256] E. Nagel, Review of von Wright [118], *Journal of Philosophy* **38** (1941) 529–31.

[257] E. Nagel, Review of von Neumann and Morgenstern [112], *Journal of Philosophy* **42** (1945) 550–4.

[258] E. Nagel, Review of Williams [115], *Journal of Philosophy* **44** (1947) 685–93.

[259] E. Nagel, Review of Reichenbach [93], *Journal of Philosophy* **47** (1950) 551–5.

[260] E. Nagel, Review of Kneale [52], *Journal of Philosophy* **47** (1950) 545–51.

[261] E. Nagel, Review of Braithwaite [12], *Philosophical Review* **66** (1957) 205–25.

[262] G. Nakhnikian, Review of Chisholm *et al.* [20], *Philosophical Quarterly* **16** (1966) 377.

[263] S. G. O'Hair, Reviews of Chisholm [19] and Scheffler [99], *Synthese* **16** (1966) 381–93.

[264] A. Pap, Review of Goodman [36], *Review of Metaphysics* **9** (1955) 285–99.

[265] H. H. Price, Review of Chisholm [18], *The British Journal for the Philosophy of Science* **10** (1959–60) 147.

[266] A. Quinton, Review of Chisholm *et al.* [20], *Philosophical Books* **6** (1965) 4.

[267] N. Rescher, Review of Popper [87], *Philosophy and Phenomenological Research* **21** (1960–61) 266–8.

[268] M. D. Resnik, Review of Hempel [40], *Philosophy and Phenomenological Research* **27** (1966–67) 139.

[269] A. Robinson, Review of Jeffreys [48], *Journal of Symbolic Logic* **29** (1964) 194.

[270] A. Robinson, Review of Jeffrey [46], *The British Journal for the Philosophy of Science* **19** (1968) 177.

[271] R. Rosthal, Review of Barker [8], *Philosophy and Phenomenological Research* **19** (1958–59) 123–4.

[272] L. J. Russell, Review of Russell [94], *Philosophy* **24** (1949) 253–60.
[273] L. J. Russell, Review of Carnap [15], *Philosophy* **28** (1953) 272–3.
[274] W. C. Salmon, Review of Jeffreys [48], *Philosophy of Science* **24** (1957) 364–6.
[275] W. C. Salmon, Review of Barker [8], *Philosophical Review* **68** (1959) 247–53.
[276] W. C. Salmon, Review of von Wright [118], *Philosophy of Science* **26** (1959) 166.
[277] W. C. Salmon, Review of Day [26], *Philosophical Review* **72** (1963) 392–6.
[278] W. C. Salmon, Review of Kyburg [55], *Philosophy of Science* **34** (1967) 283–5.
[279] F. Schick, Review of Katz [49], *Journal of Philosophy* **60** (1963) 473.
[280] F. Schick, Review of Jeffrey [46], *Journal of Philosophy* **64** (1967) 369.
[281] M. Scriven, Review of Körner [54], *The British Journal for the Philosophy of Science* **12** (1961–62) 329–42.
[282] M. Scriven, Review of Nagel [82], *Review of Metaphysics* **17** (1963–64) 403–24.
[283] J. M. Shapiro, Review of Savage [96], *Philosophy of Science* **23** (1956) 166.
[284] A. Shimony, Review of Braithwaite [12], *Review of Metaphysics* **7** (1954) 644–60.
[285] S. Silvers, Review of Skyrms [102], *Philosophy of Science* **34** (1967) 202.
[286] J. J. C. Smart, Review of Feigl and Maxwell [29], *Australasian Journal of Philosophy* **40** (1962) 110.
[287] J. J. C. Smart, Review of Feigl and Maxwell [28], *Mind* **72** (1963) 448.
[288] R. J. Spilsbury, Review of Churchman [21], *Mind* **59** (1950) 115–6.
[289] W. T. Stace, Review of Lewis [66], *Mind* **57** (1948) 71–85.
[290] J. Stannard, Review of von Mises [78], *Philosophy of Science* **26** (1959) 387–8.
[291] A. Stenner, Review of Scheffler [98], *Philosophy of Science* **33** (1966) 80–2.
[292] R. H. Stoothoff, Review of Katz [49], *Philosophical Quarterly* **15** (1965) 85.

[293] D. C. Stove, Review of Goodman [36], *Australasian Journal of Philosophy* **33** (1955) 128–32.
[294] D. C. Stove, Review of Harrod [39], *Australasian Journal of Philosophy* **36** (1958) 173–87.
[295] D. C. Stove, Review of Katz [49], *Australasian Journal of Philosophy* **41** (1963) 269.
[296) D. C. Stove, Review of Scheffler [98], *Australasian Journal of Philosophy* **43** (1965) 109.
[297] L. Svenonius, Review of Barker [8], *Journal of Symbolic Logic* **27** (1962) 122–3.
[298] S. Toulmin, Review of Carnap [16], *Mind* **62** (1953) 86–99.
[299] J. Tucker, Review of Day [26], *Philosophical Books* **2** (1961) 6.
[300] J. Tucker, Review of Madden [69], *Philosophy of Science* **28** (1961) 86–9.
[301] J. M. Vickers, Review of Martin [74], *Journal of Philosophy* **64** (1967) 193.
[302] R. H. Vincent, Review of Carnap [16], *Dialogue* **2** (1963) 97–101.
[303] G. J. Warnock, Review of Popper [87], *Mind* **69** (1960) 99–101.
[304] J. Watling, Review of Goodman [36], *Mind* **65** (1956) 267–73.
[305] A. R. White, Review of Hintikka [41], *Philosophical Quarterly* **15** (1965) 268.
[306] M. White, Review of Ayer [5], *Encounter* **8** (1957) 80–2.
[307] C. M. Whitely, Review of Wisdom [116], *Mind* **62** (1953) 113–4.
[308] F. L. Will, Review of Williams [115], *Philosophical Review* **57** (1948) 231–47.
[309] F. L. Will, Review of Kneale [52], *Philosophical Review* **63** (1954) 19–42.
[310] F. L. Will, Review of Toulmin [109], *Philosophical Review* **69** (1960) 399.
[311] D. C. Williams, Review of Reichenbach [93], *Philosophy and Phenomenological Research* **11** (1950–51) 252–7.
[312] J. H. Woodger, Review of Braithwaite [12], *Journal of Symbolic Logic* **22** (1957) 404–6.
[313] J. Yolton, Review of Chisholm [18], *Philosophy of Science* **25** (1958) 302–5.
[314] R. Young, Review of Feyerabend and Maxwell [30], *The British Journal for the Philosophy of Science* **18** (1968) 325.

III. ARTICLES

A. *Induction, Acceptance, and Rational Belief*

[315] J. Agassi, 'Corroboration Versus Induction', *The British Journal for the Philosophy of Science* **9** (1958–59) 311–7.

[316] J. Agassi, 'The Role of Corroboration in Popper's Methodology', *Australasian Journal of Philosophy* **39** (1961) 82–91.

[317] Y. Bar-Hillel, 'On Alleged Rules of Detachment in Inductive Logic', in Lakatos [58] 120–8.

[318] Y. Bar-Hillel, 'The Acceptance Syndrome', in Lakatos [58] 150–61.

[319] R. B. Braithwaite, 'The Role of Values in Scientific Inference', in Kyburg and Nagel [56] 180–93.

[320] G. Buchdahl, 'Induction and Scientific Method', *Mind* **60** (1961) 16–34.

[321] M. Bunge, 'The Place of Induction in Science', *Philosophy of Science* **27** (1960) 262–70.

[322] R. Carnap, 'On the Application of Inductive Logic', *Philosophy and Phenomenological Research* **8** (1947–48) 133–48.

[323] R. Carnap, 'The Aim of Inductive Logic', in Nagel *et al.* [83] 308–18.

[324] R. Carnap, 'On Rules of Acceptance', in Lakatos [58] 146–50.

[325] R. Carnap, 'Inductive Logic and Inductive Intuition', in Lakatos [58] 258–67.

[326] R. Carnap, 'Reply to J. Hintikka', in Lakatos [58] 312–4.

[327] G. Chatalian, 'Induction and the Problem of the External World', *Journal of Philosophy* **49** (1952) 601–7.

[328] R. Chisholm, 'Epistemic Statements and the Ethics of Belief', *Philosophy and Phenomenological Research* **16** (1955–56) 447–60.

[329] R. Chisholm, '"Appear", "Take", "Evident"', *Journal of Philosophy* **53** (1956) 722–31.

[330] R. Chisholm, 'Evidence as Justification', *Journal of Philosophy* **58** (1961) 739–48.

[331] R. Chisholm, 'Theory of Knowledge', in Chisholm *et al.* [20] 261–86.

[332] R. Ennis, 'Enumerative Induction and Best Explanation', *Journal of Philosophy* **65** (1968) 523–9.

333 H . A. Finch, 'Confirming Power of Observations Metricized for

Decisions among Hypotheses', *Philosophy of Science* **27** (1960) 293–307 (Part I); 391–404 (Part II).

[334] F. B. Fitch, 'Justification in Science', in White [114] 99–107.

[335] R. Firth, 'Ultimate Evidence', *Journal of Philosophy* **53** (1956) 732–9.

[336] R. Firth, 'Chisholm and the Ethics of Belief', *Philosophical Review* **68** (1959) 493–506.

[337] R. Firth, 'Coherence, Certainty, and Epistemic Priority', *Journal of Philosophy* **61** (1964) 547–57.

[338] R. Firth, 'The Anatomy of Certainty', *Philosophical Review* **76** (1967) 3–27.

[339] P. G. Frank, 'The Variety of Reasons for the Acceptance of Scientific Theories', in Frank [33] 3–18.

[340] I. J. Good, 'Rational Decisions', *Journal of the Royal Statistical Society*, series B, **14** (1952) 107–14.

[341] I. J. Good, 'Weight of Evidence, Corroboration, Explanatory Power, and the Utility of Experiments', *Journal of the Royal Statistical Society* **22** (1960) 319–31.

[342] I. J. Good, 'Corroboration, Explanation, Evolving Probabilities, Simplicity and a Sharpened Razor', *The British Journal for the Philosophy of Science* **19** (1968) 123–43.

[343] I. Hacking, 'One Problem about Induction', in Lakatos [58] 44–59.

[344] G. Harman, 'How Belief is Based on Inference', *Journal of Philosophy* **61** (1964) 353–9.

[345] G. Harman, 'The Inference to the Best Explanation', *Philosophical Review* **74** (1965) 88–95.

[346] G. Harman, 'Detachment, Probability, and Maximum Likelihood', *Nous* **1** (1967) 401–11.

[347] G. Harman, 'Knowledge, Inference, and Explanation', *American Philosophical Quarterly* **5** (1968) 164–73.

[348] G. Harman, 'Enumerative Induction as Inference to the Best Explanation', *Journal of Philosophy* **65** (1968) 529–33.

[349] W. Harper, (see Kyburg [370]).

[350] H. Heidelberger, 'Knowledge, Certainty, and Probability', *Inquiry* **6** (1963) 242–50.

[351] H. Heidelberger, 'Chisholm's Epistemic Principles', *Nous* **3** (1969) 73–82.

[352] C. G. Hempel, 'Studies in the Logic of Confirmation', in Hempel [40] 2–51.

[353] C. G. Hempel, 'Inductive Inconsistencies', *Synthese* **12** (1960) 439–69.

[354] C. G. Hempel, 'Deductive-Nomological vs. Statistical Explanation', in Feigl and Maxwell [28] 98–169.

[355] P. Henle, 'On the Certainty of Empirical Statements', *Journal of Philosophy* **44** (1947) 625–32.

[356] R. Hilpinen, 'Rules of Acceptance and Inductive Logic', *Acta Philosophica Fennica*, fasc. **22** (1968). (See Corrigendum, p. 226.)

[357] R. Hilpinen, (see Hintikka [359]).

[358] J. Hintikka, 'The Possibility of Rules of Acceptance', in Lakatos [58] 144–6.

[359] J. Hintikka and R. Hilpinen, 'Knowledge, Acceptance and Inductive Logic', in Hintikka and Suppes [42] 1–21.

[360] J. Hintikka and J. Pietarinen, 'Semantic Information and Inductive Logic', in Hintikka and Suppes [42] 96–112.

[361] R. C. Jeffrey, 'Valuation and Acceptance of Scientific Hypotheses', *Philosophy of Science* **23** (1956) 237–46.

[362] R. C. Jeffrey, 'Probable Knowledge', in Lakatos [58] 166–80.

[363] H. E. Kyburg, Jr., 'Probability and Rationality', *Philosophical Quarterly* **11** (1961) 3–10.

[364] H. E. Kyburg, Jr., 'Comments on Braithwaite's Paper', in Kyburg and Nagel [56] 196–9.

[365] H. E. Kyburg, Jr., 'A Further Note on Rationality and Consistency', *Journal of Philosophy* **60** (1963) 463–5.

[366] H. E. Kyburg, Jr., 'Probability, Rationality, and a Rule of Detachment', in Bar-Hillel [7] 301–10.

[367] H. E. Kyburg, Jr., 'Probability and Decision', *Philosophy of Science* **33** (1966) 250–61.

[368] H. E. Kyburg, Jr., 'The Rule of Detachment in Inductive Logic', in Lakatos [58] 98–119.

[369] H. E. Kyburg, Jr., 'Detaching Estimates', in Lakatos [58] 72–4.

[370] H. E. Kyburg, Jr. and W. Harper, 'The Jones Case', *The British Journal for the Philosophy of Science* **19** (1968) 247–51.

[371] K. Lehrer, 'Induction, Reason, and Consistency', in *Minnesota Studies in the Philosophy of Science* (forthcoming).

[372] K. Lehrer *et al.*, 'Reason and Evidence: An Unsolved Problem', Ratio **9** (1967) 38–48.

[373] I. Levi, 'Must the Scientist Make Value Judgments?', *Journal of Philosophy* **57** (1960) 345–57.

[374] I. Levi, 'Decision Theory and Confirmation', *Journal of Philosophy* **58** (1961) 614–25.

[375] I. Levi, 'On the Seriousness of Mistakes', *Philosophy of Science* **29** (1962) 47–65.

[376] I. Levi, 'Corroboration and Rules of Acceptance', *The British Journal for the Philosophy of Science* **13** (1962–63) 307–13.

[377] I. Levi, 'Belief and Action', *The Monist* **48** (1964) 306–16.

[378] I. Levi, 'Deductive Cogency in Inductive Inference', *Journal of Philosophy* **62** (1965) 68–77.

[379] I. Levi, 'Utility and Acceptance of Hypotheses', in Morgenbesser [79] 115–24.

[380] I. Levi, 'Information and Inference', *Synthese* **17** (1967) 369–91.

[381] I. Levi, 'Probability Kinematics', *The British Journal for the Philosophy of Science* **18** (1967) 197–209.

[382] N. Malcolm, 'Certainty and Empirical Statements', *Mind* **51** (1942) 18–46.

[383] J. Pietarinen, (see Hintikka [360]).

[384] K. R. Popper, 'Probability Magic or Knowledge Out of Ignorance', *Dialectica* **2** (1957) 354–73.

[385] K. R. Popper, 'On Rules of Detachment and So-called Inductive Logic', in Lakatos [58] 130–9.

[386] K. R. Popper, 'Theories, Experience, and Probabilistic Intuitions', in Lakatos [58] 285–303.

[387] F. P. Ramsey, 'Truth and Probability', in Ramsey [92] 156–98.

[388] F. P. Ramsey, 'Further Considerations', in Ramsey [92] 199–211 (A: Reasonable degree of belief. B: Statistics. C: Chance).

[389] N. Rescher, 'Theory of Evidence', *Philosophy of Science* **25** (1958) 83–94.

[390] N. Rescher, 'Plausible Implication', *Analysis* **21** (1960–61) 128–35.

[391] N. Rescher, 'Non-Deductive Rules of Inference and Problems in the Analysis of Inductive Reasoning', *Synthese* **13** (1961) 242–51.

[392] R. Roelofs, (see Lehrer [372]).

[393] R. S. Rudner, 'The Scientist Qua Scientist Makes Value Judgments', *Philosophy of Science* **20** (1953) 1–6.

[394] R. S. Rudner, 'Value Judgments in the Acceptance of Theories', in Frank [33] 24–8.

[395] W. C. Salmon, 'The Justification of Inductive Rules in Inference', in Lakatos [58] 24–43.

[396] W. C. Salmon, 'Who Needs Inductive Acceptance Rules', in Lakatos [58] 24–43 and 139–44.

[397] I. Scheffler, 'On Justification and Commitment', *Journal of Philosophy* **51** (1954) 180–90.

[398] F. Schick, 'Rationality and Consistency', *Journal of Philosophy* **60** (1963) 5–19.

[399] F. Schick, 'Consistency', *Philosophical Review* **75** (1966) 467–95.

[400] A. Shimony, 'Coherence and the Axioms of Confirmation', *Journal of Symbolic Logic* **20** (1955) 1–28.

[401] H. Smokler, 'Goodman's Paradox and the Problem of Rules of Acceptance', *American Philosophical Quarterly* **3** (1966) 71.

[402] P. Suppes, 'The Philosophical Relevance of Decision Theory', *Journal of Philosophy* **58** (1961) 605–14.

[403] P. Suppes, 'Probabilistic Inference and the Concept of Total Evidence', in Hintikka and Suppes [42] 49–65.

[404] P. Suppes, 'Rational Changes of Belief', in Lakatos [58] 186–8.

[405] M. Swain, 'Schick on Consistency', *Philosophical Studies* (forthcoming).

[406] M. Swain, (see Lehrer [372]).

[407] J. W. Tukey, 'Conclusions versus Decisions', *Technometrics* **2** (1960) 423–33.

[408] F. Will, 'The Preferability of Probable Beliefs', *Journal of Philosophy* **62** (1965) 57–67.

[409] P. M. Williams, 'The Structure of Acceptance and its Evidential Basis', *The British Journal for the Philosophy of Science* **19** (1969) 325–44.

[410] G. H. von Wright, 'Remarks on the Epistemology of Subjective Probability' in Nagel *et al.* [83] 330–9.

B. *Related Topics*

[411] P. Achinstein, 'Variety and Analogy in Confirmation Theory', *Philosophy of Science* **30** (1963) 207–21.

[412] P. Achinstein, (see Barker [432]).

[413] R. Ackerman, 'Inductive Simplicity', *Philosophy of Science* **28** (1961) 152–61.

[414] R. Ackerman, 'Some Remarks on Kyburg's Modest Proposal', *Philosophical Review* **71** (1962) 236–40.

[415] J. Agassi, 'The Mystery of the Ravens', *Philosophy of Science* **33** (1966) 395–402.

[416] H. G. Alexander, 'The Paradoxes of Confirmation', *The British Journal for the Philosophy of Science* **9** (1958–59) 227–33.

[417] H. G. Alexander, 'The Paradoxes of Confirmation – A Reply to Dr. Agassi', *The British Journal for the Philosophy of Science* **10** (1959–60) 229–34.

[418] J. E. J. Altham, 'A Note on Goodman's Paradox', *The British Journal for the Philosophy of Science* **19** (1968) 257.

[419] R. B. Angel, 'Explanation and Prediction, a Plea for Reason', *Philosophy of Science* **34** (1967) 276–82.

[420] F. J. Anscombe, 'Mr. Kneale on Probability and Induction', *Mind* **60** (1951) 299–309.

[421] K. J. Arrow, 'Alternative Approaches to the Theory of Choice in Risk-Taking Situations', *Econometrica* **19** (1951) 404–37.

[422] A. J. Ayer, 'On the Probability of Particular Events', *Revue Internationale de Philosophie* **15** (1961) 366–75.

[423] P. Baille, 'That Confirmation may yet be a Probability', *The British Journal for the Philosophy of Science* **20** (1969) 41–51.

[424] Y. Bar-Hillel, 'A Note on Comparative Inductive Logic', *The British Journal for the Philosophy of Science* **3** (1952–53) 308–10.

[425] Y. Bar-Hillel, 'An Examination of Information Theory', *Philosophy of Science* **22** (1955) 86–105.

[426] Y. Bar-Hillel, 'Comments on "Degree of Confirmation" by Professor K. R. Popper', *The British Journal for the Philosophy of Science* **6** (1955–56) 155–7.

[427] Y. Bar-Hillel, 'Further Comments on Probability and Confirmation', *The British Journal for the Philosophy of Science* **7** (1956–57) 245–8.

[428] Y. Bar-Hillel and R. Carnap, 'Semantic Information', *The British Journal for the Philosophy of Science* **4** (1953) 145–57.

[429] Y. Bar-Hillel, (see Carnap [466]).

[430] S. Barker, 'The Role of Simplicity in Explanation', in Feigl and Maxwell [29] 265–74.

[431] S. Barker, 'On Simplicity in Empirical Hypotheses', *Philosophy of Science* **28** (1961) 162–71.

[432] S. Barker and P. Achinstein, 'On the New Riddle of Induction', *Philosophical Review* **69** (1960) 511–22.

[433] W. W. Bartley, III, 'A Note on Barker's Discussion of Popper's Theory of Corroboration', *Philosophical Studies* **12** (1961) 5–10.

[434] W. W. Bartley III, 'Goodman's Paradox: A Simpleminded Solution', *Philosophical Studies* **19** (1968) 85–7.

[435] W. H. Baumer, 'Confirmation Still without Paradoxes', *The British Journal for the Philosophy of Science* **19** (1968) 57–63.

[436] T. Bayes, 'An Essay Towards Solving a Problem in the Doctrine of Chances (with and Introductory Letter, Footnotes, and an Appendix by R. Price)', *Philosophical Transactions of the Royal Society* **53** (1763) 370–418.

[437] Jonathan F. Bennett, 'Some Aspects of Probability and Induction (I)', *The British Journal for the Philosophy of Science* **7** (1956–57) 220–30.

[438] Jonathan F. Bennett, 'Some Aspects of Probability and Induction (II)', *The British Journal for the Philosophy of Science* **7** (1956–57) 316–22.

[439] C. W. Berenda, 'On Verifiability, Simplicity, and Equivalence', *Philosophy of Science* **19** (1952–53) 70–6.

[440] Gustav Bergmann, 'Some Comments on Carnap's Logic of Induction', *Philosophy of Science* **13** (1946) 71–8.

[441] D. E. Berlyne, 'Comments on Salmon's Paper', in Kyburg and Nagel [56] 42–4.

[442] E. Bolker, 'A Simultaneous Axiomatization of Utility and Subjective Probability', *Philosophy of Science* **34** (1967) 333–40.

[443] R. B. Braithwaite, 'On Unknown Probabilities', in Körner [54] 3–11.

[444] B. A. Brody, 'Confirmation and Explanation', *Journal of Philosophy* **65** (1968) 282–99.

[445] D. G. Brown, 'The Nature of Inference', *Philosophical Review* **64** (1955) 351–69.

[446] Jeffrey Bub and Michale Radner, 'Miller's Paradox of Information',

The British Journal for the Philosophy of Science **19** (1968) 63–7.

[447] M. Bunge, 'The Weight of Simplicity in the Construction and Assaying of Scientific Theories', *Philosophy of Science* **28** (1961) 120–49.

[448] A. W. Burks, 'Pierce's Theory of Abduction', *Philosophy of Science* **13** (1946) 301–6.

[449] A. W. Burks, 'Reichenbach's Theory of Probability and Induction', *Review of Metaphysics* **4** (1951) 377–93.

[450] A. W. Burks, 'The Presupposition Theory of Induction', *Philosophy of Science* **20** (1953) 177–97.

[451] A. W. Burks, 'On the Presuppositions of Induction', *Review of Metaphysics* **8** (1954–55) 574–611.

[452] A. W. Burks, 'On the Significance of Carnap's System of Inductive Logic for the Philosophy of Induction', in Schilpp [100] 739–59.

[453] J. Canfield and Keith Lehrer, 'A Note in Prediction and Deduction', *Philosophy of Science* **28** (1961) 204–8.

[454] R. Carnap, 'On Inductive Logic', *Philosophy of Science* **12** (1945) 72–97.

[455] R. Carnap, 'Reply to Nelson Goodman', *Philosophy and Phenomenological Review* **8** (1947) 461–2.

[456] R. Carnap, 'The Problem of Relations in Inductive Logic', *Philosophical Studies* **2** (1951) 75–80.

[457] R. Carnap, 'On the Comparative Concept of Confirmation', *The British Journal for the Philosophy of Science* **3** (1952) 65–73.

[458] R. Carnap, 'Remarks to Kemeny's Paper', *Philosophy and Phenomenological Research* **13** (1952–53) 375–6.

[459] R. Carnap, 'What is Probability?', *Scientific American* **189** (1953) 128–38.

[460] R. Carnap, 'Remarks on Popper's Note on Content and Degree of Confirmation', *The British Journal for the Philosophy of Science* **7** (1956–57) 243–4.

[461] R. Carnap, 'Variety, Analogy and Periodicity in Inductive Logic', *Philosophy of Science* **30** (1963) 222-7.

[462] R. Carnap, 'Replies and Systematic Expositions', in Schilpp [100] 966–98.

[463] R. Carnap, 'Probability and Content Measure', in Feyerabend and Maxwell [30] 248–60.

[464] R. Carnap, 'Aspects of Hempel's Philosophy of Science', *Review of Metaphysics* **20** (1967) 690–710.
[465] R. Carnap, (see Bar-Hillel [428]).
[466] R. Carnap and Y. Bar-Hillel, *An Outline of the Theory of Semantic Information*, Res. Lab. of Electronics, M.I.T., Rept. No. 247, Boston 1952.
[467] N. Champawat, (see Saunders [671]).
[468] Michael Clark, 'Knowledge and Grounds: A Comment on Mr. Gettier's Paper', *Analysis* **24** (1963) 46–8.
[469] L. J. Cohen, 'Claims to Knowledge', *Proceedings of the Aristotelian Society, Supp.* **36** (1962) 33–50.
[470] L. J. Cohen, 'What has Confirmation to do with Probabilities?', *Mind* **75** (1966) 463–81.
[471] A. H. Copeland, 'Statistical Induction and the Foundations of Probability', *Theoria* **28** (1962) 27–44 and 87–109.
[472] Jared Darlington, 'On the Confirmation of Laws', *Philosophy of Science* **26** (1959) 14–24.
[473] Jared Darlington, 'Reply to Linhart', *Philosophy of Science* **26** (1959) 363.
[474] R. Das, 'Induction and Non-Instantial Hypothesis', *The British Journal for the Philosophy of Science* **8** (1957–58) 317–25.
[475] D. Davidson and P. Suppes, 'A Finitistic Axiomatization of Subjective Probability and Utility', *Econometrica* **24** (1956) 264–75.
[476] R. Eberle *et al.*, 'Hempel and Oppenheim on Explanation', *Philosophy of Science* **28** (1961) 418–28.
[477] Haskell Fain, 'The Very Thought of Grue', *Philosophical Review* **74** (1967) 61–73.
[478] H. Feigl, 'Confirmability and Confirmation', *Revue Internationale de Philosophie* **5** (1951) 268–79.
[479] Lewis S. Feuer, 'The Principle of Simplicity', *Philosophy of Science* **24** (1957) 109–22.
[480] Paul K. Feyerabend, 'A Note on Two Problems of Induction', *The British Journal for the Philosophy of Science* **19** (1968) 251–3.
[481] B. de Finetti, 'La vrai et la probable', *Dialectica* **3** (1949) 78–92.
[482] B. de Finetti, 'Foundations of Probability', in Klibansky [51] 140–7.
[483] B. de Finetti, 'Expérience et théorie dans l'élaboration et dans l'application d'une doctrine scientifique' (Troisièmes Entretiens

de Zurich, 1951), *Revue de Métaphysique* **60** (1955) 264–86.
[484] B. de Finetti, 'Foresight: Its Logical Laws, Its Subjective Sources', in Kyburg and Smokler [57] 97–158.
[485] D. Föllesdal, 'Knowledge, Identity, and Existence', *Theoria* **33** (1967) 1–27.
[486] John Freund, 'On the Problem of Confirmation', *Methodos* **3** (1951) 33–42.
[487] John Freund, 'On the Confirmation of Scientific Theories', *Philosophy of Science* **17** (1950) 87–94.
[488] M. Friedman and L. J. Savage, 'The Utility Analysis of Choices Involving Risk', *Journal of Political Economy* **56** (1948) 279–304.
[489] E. Gettier, 'Is Justified True Belief Knowledge?', *Analysis* **23** (1963) 121–3. Reprinted in Griffiths [37] 144–6.
[490] L. Gibson, 'On "Ravens and Relevance" and a Likelihood Solution of the Paradox of Confirmation', *The British Journal for the Philosophy of Science* **20** (1969) 75–80.
[491] A. Goldman, 'A Causal Theory of Knowing', *Journal of Philosophy* **64** (1967) 357–72.
[492] I. J. Good, 'Kinds of Probability', *Science* **129** (1959) 443–7.
[493] I. J. Good,'The Paradox of Confirmation (II)', *The British Journal for the Philosophy of Science* **11** (1960–61) 145–9.
[494] I. J. Good, 'On the Principle of Total Evidence', *The British Journal for the Philosophy of Science* **17** (1967) 319–21.
[495] I. J. Good, 'The White Shoe is a Red Herring (Paradox of Confirmation)', *The British Journal for the Philosophy of Science* **17** (1967) 322.
[496] I. J. Good, 'The White Shoe *qua* Herring is Pink', *The British Journal for the Philosophy of Science* **19** (1968) 156–7.
[497] I. J. Good, 'Subjective Probability as a Measure of a Non-measureable Set', in Nagel *et al.* [83] 319–29.
[498] N. Goodman, 'A Query on Confirmation', *Journal of Philosophy* **43** (1946) 383–5.
[499] N. Goodman, 'On Infirmities of Confirmation Theory', *Philosophy and Phenomenological Research* **8** (1947–48) 149–51.
[500] N. Goodman, 'New Notes on Simplicity', *Journal of Symbolic Logic* **17** (1952) 189–91.

[501] N. Goodman, 'Axiomatic Measurement of Simplicity', *Journal of Philosophy* **52** (1955) 709–22.

[502] N. Goodman, 'Recent Developments in the Theory of Simplicity'. *Philosophy and Phenomenological Research* **19** (1958–59) 429–46.

[503] N. Goodman, 'The Test of Simplicity', *Science* **128** (1958) 1064-9.

[504] N. Goodman, 'Positionality and Pictures', *Philosophical Review* **69** (1960) 523–5.

[505] N. Goodman, 'Safety, Strength, Simplicity', *Philosophy of Science* **28** (1961) 150–1.

[506] N. Goodman, 'Comments', *Journal of Philosophy* **63** (1966) 328–31.

[507] Richard E. Grandy, 'Some Comments on Confirmation and Selective Confirmation', *Philosophical Studies* **18** (1967) 19–24.

[508] I. Hacking, 'Slightly More Realistic Personal Probability', *Philosophy of Science* **34** (1967) 311–25.

[509] I. Hacking, 'Discussion: On Falling Short of Strict Coherence', *Philosophy of Science* **35** (1968) 284–6.

[510] C. L. Hamblin, 'The Modal "Probably"', *Mind* **68** (1959) 234–40.

[511] Marsha Hanen, 'Goodman, Wallace, and the Equivalence Condition', *Journal of Philosophy* **64** (1967) 271-80.

[512] G. Harman, 'Lehrer on Knowledge', *Journal of Philosophy* **63** (1966) 241–7.

[513] G. Harman, 'Unger on Knowledge', *Journal of Philosophy* **64** (1967) 390–5.

[514] R. Harré, 'Simplicity as a Criterion of Induction', *Philosophy* **34** (1959) 229–34.

[515] R. F. Harrod, 'The General Structure of Inductive Argument', *Proceedings of the Aristotelian Society* **52** (1951–52) 41–56.

[516] John C. Harsanyi, 'Popper's Improbability Criterion for the Choice of Scientific Hypotheses', *Philosophy* **35** (1960) 332–40.

[517] William H. Hay, 'Professor Carnap and Probability', *Philosophy of Science* **19** (1952) 170–7.

[518] F. Hayek, 'Degrees of Explanation', *The British Journal for the Philosophy of Science* **6** (1955–56) 209–25.

[519] H. Heidelberger, 'On Defining Epistemic Terms', *Journal of Philosophy* **60** (1963) 344–8.

[520] H. Heidelberger, 'An Undefined Epistemic Term', *Mind* **75** (1966) 420-1.

[521] Olaf Helmer and Paul Oppenheim, 'A Syntactical Definition of Probability and of Degree of Confirmation', *Journal of Symbolic Logic* **10** (1945) 25–60.

[522] C. G. Hempel, 'A Purely Syntactical Definition of Confirmation', *Journal of Symbolic Logic* **8** (1943) 122–43.

[523] C. G. Hempel, 'Empirical Statements and Falsifiability', *Philosophy* **33** (1958) 342–8.

[524] C. G. Hempel, 'The Theoretician's Dilemma', in Hempel [40] 173–226.

[525] C. G. Hempel, 'Aspects of Scientific Explanation', in Hempel [40] 331–496.

[526] C. G. Hempel, 'The White Shoe, No Red Herring', *The British Journal for the Philosophy of Science* **18** (1967) 239–40.

[527] C. G. Hempel, 'Maximal Specificity and Lawlikeness in Probabilistic Explanation', *Philosophy of Science* **35** (1968) 116–33.

[528] C. G. Hempel, 'Discussion: On a Claim by Skyrms Concerning Lawlikeness and Confirmation', *Philosophy of Science* **35** (1968) 274–8.

[529] C. G. Hempel and Paul Oppenheim, 'A Definition of "Degree of Confirmation"', *Philosophy of Science* **12** (1945) 98–115.

[530] C. G. Hempel and Paul Oppenheim, 'Studies in the Logic of Explanation', *Philosophy of Science* **15** (1948) 135–75.

[531] M. Hesse, 'Ramifications of "Grue"', *The British Journal for the Philosophy of Science* **20** (1969) 13–25.

[532] Donald Hillman, "The Measurement of Simplicity', *Philosophy of Science* **29** (1962) 225–52.

[533] Risto Hilpinen, 'On Inductive Generalization in Binary Functional Calculus' (a paper read at the 3rd International Congress for Logic, Methodology, and Philosophy of Science, Amsterdam 1967).

[534] K. J. Hintikka, 'On a Combined System of Inductive Logic', *Acta Philosophica Fennica* **18** (1965) 21–30.

[535] K. J. Hintikka, 'Toward a Theory of Inductive Generalization', in Bar-Hillel [7] 274–88.

[536] K. J. Hintikka, 'Induction by Enumeration and Induction by Elimination', in Lakatos [58] 191–216.

[537] K. J. Hintikka, 'A Two-Dimensional Continuum of Inductive Methods', in Hintikka and Suppes [42] 113–32.

[538] K. J. Hintikka, '"Knowing Oneself" and Other Problems in Epistemic Logic', *Theoria* **32** (1966) 3–13.

[539] K. J. Hintikka, 'Individuals, Possible Worlds, and Epistemic Logic', *Nous* **1** (1967) 33–61.

[540] K. J. Hintikka, 'Existence and Identity in Epistemic Contexts: A Comment on Føllesdal's Paper', *Theoria* **33** (1967) 138–47.

[541] K. J. Hintikka, 'Epistemic Logic and the Methods of Philosophical Analysis', *Australasian Journal of Philosophy* **46** (1968) 37–51.

[542] K. J. Hintikka, 'The Varieties of Information and Scientific Explanation', in Van Rootselaar and Staal [110] 311–31.

[543] K. J. Hintikka, 'Partially Transparent Senses of Knowing', *Philosophical Studies* **20** (1969) 4–8.

[544] C. A. Hooker, 'Goodman, "Grue", and Hempel', *Philosophy of Science* **35** (1968) 232–47.

[545] C. A. Hooker and D. Stove, 'Relevance and the Ravens', *The British Journal for the Philosophy of Science* **18** (1968) 305–15.

[546] James Hullett and Robert Schwartz, 'Grue: Some Remarks', *Journal of Philosophy* **64** (1967) 259–71.

[547] M. Hummerton, 'Bayesian Statistics and Popper's Epistemology', *Mind* **77** (1968) 109–17.

[548] R. C. Jeffrey, 'Comments on Leblanc Paper', in Kyburg and Nagel [56] 18–21.

[549] R. C. Jeffrey, 'Popper on the Rule of Succession', *Mind* **73** (1964) 129.

[550] R. C. Jeffrey, 'Goodman's Query', *Journal of Philosophy* **63** (1966) 281–8.

[551] Charles Jordan, 'On Daniel Bernoulli's "Moral Expectation" and on a New Conception of Expectation', *American Mathematical Monthly* **31** (1924) 183–90.

[552] Howard Kahane, 'Baumer on the Confirmation Paradox', *The British Journal for the Philosophy of Science* **18** (1967) 52–6.

[553] D. Kaplan (see Eberle [476]).

[554] Felix Kaufmann, 'Verification, Meaning, and Truth', *Philosophy and Phenomenological Research* **4** (1943) 267–86.

[555] G. B. Keene, 'Randomness', *Proceedings of the Aristotelian Society, Supp.* **31** (1957) 151–60.

[556] G. B. Keene, 'Mill's Method of Hypothesis', *Filosophia, Supp.* **13** (1962) 595–8.

[557] E. C. Kemble, 'Is the Frequency Theory of Probability Adequate for all Scientific Purposes?', *American Journal of Physics* **10** (1942) 6–16.

[558] J. G. Kemeny, 'A Contribution to Inductive Logic', *Philosophy and Phenomenological Research* **13** (1952–53) 371–4.

[559] J. G. Kemeny, 'The Use of Simplicity in Induction', *Philosophical Review* **62** (1953) 391–408.

[560] J. G. Kemeny, 'Two Measures of Complexity', *Journal of Philosophy* **52** (1955) 722–33.

[561] J. G. Kemeny, 'Fair Bets and Inductive Probabilities', *Journal of Symbolic Logic* **20** (1955) 263–73.

[562] J. G. Kemeny, 'Carnap's Theory of Probability and Induction' in Schilpp [100] 711–38.

[563] J. G. Kemeny and Paul Oppenheim, 'Degree of Factual Support', *Philosophy of Science* **19** (1952) 307–24.

[564] W. Kneale, 'Probability and Induction', *Mind* **60** (1951) 310–7.

[565] W. Kneale, 'Some Aspects of Probability and Induction: A Reply to Mr. Bennett', *The British Journal for the Philosophy of Science* **8** (1957–58) 57–63.

[566] Henry E. Kyburg, Jr., 'The Justification of Induction', *Journal of Philosophy* **53** (1956) 392–400.

[567] Henry E. Kyburg, Jr., 'R. B. Braithwaite on Probability and Induction', *The British Journal for the Philosophy of Science* **9** (1958–59) 203–20.

[568] Henry E. Kyburg, Jr., 'Demonstrative Induction', *Philosophy and Phenomenological Research* **21** (1960–61) 80–92.

[569] Henry E. Kyburg, Jr., 'A Modest Proposal Concerning Simplicity', *The Philosophical Review* **70** (1961) 390–5.

[570] Henry E. Kyburg Jr., 'Probability and Randomness', *Theoria* **29** (1963) 27–55.

[571] Henry E. Kyburg, Jr., 'Recent Work in Inductive Logic', *American Philosophical Quarterly* **1** (1964) 249–87.

[572] Henry E. Kyburg, Jr., 'Logical and Fiducial Probability', *Bulletin of the International Statistical Institute*, Proceedings of the 34th session. Ottawa 1963.

[573] Henry E. Kyburg, Jr., 'Bets and Beliefs', *American Philosophical Quarterly* **5** (1968) 54–63.

[574] Hugues Leblanc, 'Evidence logique et degrés de confirmation', *Revue Philosophique de Louvain* **52** (1954) 619–25.

[575] Hugues Leblanc, 'On So-Called Degrees of Confirmation', *The British Journal for the Philosophy of Science* **10** (1959-60) 312–5.

[576] Hugues Leblanc, 'On Chances and Estimated Chances of Being True', *Revue Philosophique de Louvain* **57** (1959) 225–39.

[577] Hugues Leblanc, 'The Problem of the Confirmation of Laws', *Philosophical Studies* **12** (1961) 81–4.

[578] Hugues Leblanc, 'That Positive Instances Are No Help', *Journal of Philosophy* **60** (1963) 453–62.

[579] Hugues Leblanc, 'A Revised Version of Goodman's Paradox on Confirmation', *Philosophical Studies* **14** (1963) 49–51.

[580] Hugues Leblanc, 'Statistical and Inductive Probabilities', in Kyburg and Nagel [56] 3–16.

[581] R. Sherman Lehman, 'On Confirmation and Rational Betting', *Journal of Symbolic Logic* **20** (1955) 251–62.

[582] K. Lehrer, 'Descriptive Completeness and Inductive Methods', *Journal of Symbolic Logic* **28** (1963) 157–60.

[583] K. Lehrer, 'Knowledge and Probability', *Journal of Philosophy* **61** (1964) 368–72.

[584] K. Lehrer, 'Knowledge, Truth, and Evidence', *Analysis* **25** (1965) 168–75.

[585] K. Lehrer, 'Letter: On Knowledge and Probability', *Journal of Philosophy* **62** (1965) 67–8.

[586] K. Lehrer, 'Belief and Knowledge', *Philosophical Review* **77** (1968) 491–9.

[587] K. Lehrer, (see Canfield [453]).

[588] K. Lehrer and T. Paxson, 'Knowledge: Undefeated Justified True Belief', *Journal of Philosophy* **66** (1969) 225–37.

[589] John W. Lenz, 'Carnap on Defining "Degrees of Confirmation"', *Philosophy of Science* **23** (1956) 230–6.

[590] Isaac Levi, 'Hacking, Salmon on Induction', *Journal of Philosophy* **62** (1965) 481–7.

[591] Isaac Levi, 'On Potential Surprise', *Ratio* **8** (1966) 107–29.

[592] C. I. Lewis, 'The Given Element in Empirical Knowledge', *Philosophical Review* **61** (1952) 168–75.

[593] D. V. Lindley, 'Statistical Inference', *Journal of the Royal Statistical Society*, series B, **15** (1953) 30–76.

[594] H. Linhart, 'Darlington's "On the Confirmation of Laws"', *Philosophy of Science* **26** (1949) 362.

[595] L. Linski, 'On Interpreting Doxastic Logic', *Journal of Philosophy* **65** (1968) 500–2.

[596] J. L. Mackie, 'The Paradox of Confirmation', *The British Journal for the Philosophy of Science* **13** (1962–63) 265–77.

[597] J. L. Mackie, 'Miller's So-Called Paradox of Information', *The British Journal for the Philosophy of Science* **17** (1966) 144–7.

[598] J. L. Mackie, 'The Relevance Criterion of Confirmation', *The British Journal for the Philosophy of Science* **20** (1969) 27–40.

[599] E. H. Madden, 'The Riddle of Induction', *Journal of Philosophy* **55** (1958) 705–18.

[600] W. I. Matson, 'Against Induction and Empiricism', *Proceedings of the Aristotelian Society* **62** (1961–62) 143–58.

[601] Gerald J. Massey, 'Hempel's Criterion of Maximal Specificity', *Philosophical Studies* **19** (1968) 43–7.

[602] Donald McQueen, 'Inference and Novelty', *Analysis* **28** (1967) 49–55.

[603] D. H. Mellor, 'Inexactness and Explanation', *Philosophy of Science* **33** (1966) 345–59.

[604] D. H. Mellor, 'Connectivity, Chance, and Ignorance', *The British Journal for the Philosophy of Science* **18** (1967) 235–8.

[605] D. H. Mellor, 'Imprecision and Explanation', *Philosophy of Science* **34** (1967) 1–9.

[606] W. W. Mellor, 'Knowing, Believing, and Behaving', *Mind* **76** (1967) 327–45.

[607] David Miller, 'A Paradox of Information', *The British Journal for the Philosophy of Science* **17** (1966) 59–61.

[608] David Miller, 'On a So-Called Paradox: A Reply to Professor J. L. Mackie', *The British Journal for the Philosophy of Science* **17** (1967) 147–9.

[609] David Miller, 'The Straight and Narrow Rule of Induction, a Reply to Dr. Bub and Mr. Radner', *The British Journal for the Philosophy of Science* **19** (1968) 145–52.

[610] Dickinson S. Miller, 'Professor Donald Williams versus Hume', *Journal of Philosophy* **44** (1947) 673–84.

[611] R. Montague, (see Eberle [476]).
[612] S. Morgenbesser, 'Goodman on the Raven', *Journal of Philosophy* **59** (1962) 493–5.
[613] E. Nagel, 'Carnap's Theory of Induction', in Schilpp [100] 785–825.
[614] E. Nagel, 'Probability and non-Demonstrative Inference', *Philosophy and Phenomenological Research* **5** (1945) 485–507.
[615] J. Neyman, 'The Problem of Inductive Inference', *Communications on Pure and Applied Mathematics* **8** (1955) 13–45.
[616] J. Neyman and E. S. Pearson, 'The Testing of Statistical Hypotheses in Relation to Probabilities a priori', *Proceedings of the Cambridge Philosophical Society* **29** (1932–33) 492–510.
[617] J. Neyman and E. S. Pearson, 'On the Problem of the Most Efficient Tests of Statistical Hypotheses', *Philosophical Transactions of the Royal Society*, series A **231** (1933) 289–337.
[618] J. Neyman and E. S. Pearson, 'Contributions to the Theory of Testing Statistical Hypotheses', *Statistical Research Memoirs* I (1936) 1–37; II (1938) 25–57.
[619] Stefan Nowak, 'Some Problems of Causal Interpretation of Statistical Relationships', *Philosophy of Science* **27** (1960) 23–38.
[620] John O'Connor, 'Differential Properties and Goodman's Riddle', *Analysis* **28** (1967) 59.
[621] P. Oppenheim, (see Helmer [521]).
[622] P. Oppenheim, (see Hempel [529] and [530]).
[623] P. Oppenheim, (see Kemeny [563]).
[624] Edward J. O'Toole, 'A Note on Probability', *Philosophical Studies* (Dublin) **11** (1961–62) 112–27.
[625] C. Pailthorpe, 'Hintikka and Knowing that One Knows', *Journal of Philosophy* **64** (1967) 487–500.
[626] J. A. Passmore, 'Popper's Account of Scientific Method', *Philosophy* **35** (1960) 326–31.
[627] E. S. Pearson, (see Neyman [616], [617], [618]).
[628] William C. Pettijohn, 'Salmon on "the Short Run"', *Philosophy of Science* **23** (1956) 149.
[629] John L. Pollock, 'Chisholm's Definition of Knowledge', *Philosophical Studies* **19** (1968) 72–6.
[630] K. R. Popper, 'Degree of Confirmation', *The British Journal for the Philosophy of Science* **5** (1954–55) 143–9.

[631] K. R. Popper, '"Content" and "Degree of Confirmation": A Reply to Dr. Bar-Hillel', *The British Journal for the Philosophy of Science* **6** (1955–56) 157–63.

[632] K. R. Popper, 'Adequacy and Consistency: A Second Reply to Dr. Bar-Hillel', *The British Journal for the Philosophy of Science* **7** (1956–57) 249–56.

[633] K. R. Popper, 'A Second Note on Degree of Confirmation', *The British Journal for the Philosophy of Science* **7** (1956–57) 350–3. Reprinted in appendix *ix of Popper [87].

[634] K. R. Popper, 'Reply to Professor Carnap', *The British Journal for the Philosophy of Science* **7** (1956–57) 244–5.

[635] K. R. Popper, 'The Propensity Interpretation of the Calculus of Probability and the Quantum Theory', in Korner [54] 65–70.

[636] K. R. Popper, 'A Third Note on Degree of Corroboration or Confirmation', *The British Journal for the Philosophy of Science* **8** (1957–58) 294–302. Reprinted in appendix *ix of Popper [87].

[637] K. R. Popper, 'Corroboration by Empirical Tests', *The British Journal for the Philosophy of Science* **10** (1959–60) 315–8.

[638] K. R. Popper, 'Probabilistic Independence and Corroboration by Empirical Tests', *The British Journal for the Philosophy of Science* **10** (1959–60) 315–8.

[639] K. R. Popper, 'The Propensity Interpretation of Probability', *The British Journal for the Philosophy of Science* **10** (1959–60) 25–42.

[640] K. R. Popper, 'On the Sources of Knowledge and Ignorance', in Popper [88] 23–30.

[641] K. R. Popper, 'Truth, Rationality, and the Growth of Scientific Knowledge', in Popper [88] 215–50.

[642] K. R. Popper, 'On Carnap's Version of LaPlace's Rule of Succession', *Mind* **71** (1962) 69–73.

[643] K. R. Popper, 'A Comment on Miller's New Paradox of Information', *The British Journal for the Philosophy of Science* **17** (1966) 61–9.

[644] K. R. Popper, 'The Mysteries of Udolpho: A Reply to Professors Jeffrey and Bar-Hillel', *Mind* **76** (1967) 103–10.

[645] K. R. Popper, 'Three Views concerning Human Knowledge', in Popper [88] 97–119.

[646] H. R. Post, 'Simplicity in Scientific Theories', *The British Journal for the Philosophy of Science* **2** (1960–61) 32–41.
[647] Hilary Putnam, 'A Definition of Degree of Difference for Very Rich Languages', *Philosophy of Science* **23** (1956) 58–62.
[648] Hilary Putnam, '"Degree of Confirmation" and Inductive Logic', in Schilpp [100] 761–83.
[649] Michael Radner, (see Bub [446]).
[650] Stanislaw Rainko, 'Diachronic Epistemology: Outline of Problems', *Studia Filozoficzne* **48** (1967) 3–42.
[651] Karl Reach, 'The Foundations of our Knowledge', *Synthese* **5** (1946) 83–6.
[652] N. Rescher, 'The Problem of a Logical Theory of Belief Statements', *Philosophy of Science* **27** (1960) 95–8.
[653] N. Rescher, 'The Concept of Randomness', *Theoria* **27** (1961) 1–11.
[654] N. Rescher and F. Brian Skyrms, 'A Methodological Problem in the Evaluation of Explanation', *Nous* **2** (1968) 121–9.
[655] H. Robbins, 'A New Approach to a Classical Statistical Decision Problem', in Kyburg and Nagel [56] 101–10.
[656] Bernard Rosen, 'Chisholm on Knowledge and Principles', *Mind* **77** (1968) 411–6.
[657] W. Rozeboom, 'Why I Know So Much More Than You Do', *American Philosophical Quarterly* **4** (1967) 257–68.
[658] W. Rozeboom, 'New Dimensions of Confirmation Theory', *Philosophy of Science* **35** (1968) 134–55.
[659] W. Rozeboom, 'New Mysteries for Old: the Transfiguration of Miller's Paradox', *The British Journal for the Philosophy of Science* **19** (1969) 345–53.
[660] R. S. Rudner, 'An Introduction to Simplicity', *Philosophy of Science* **28** (1961) 109–19.
[661] Wesley Salmon, 'The Frequency Interpretation and Antecedent Probabilities', *Philosophical Studies* **4** (1953) 44–8.
[662] Wesley Salmon, 'The Uniformity of Nature', *Philosophy and Phenomenological Research* **14** (1953–54) 39–48.
[663] Wesley Salmon, 'The Short Run', *Philosophy of Science* **22** (1955) 214–21.
[664] Wesley Salmon, 'Regular Rules of Induction', *Philosophical Review* **65** (1956) 385–8.

[665] Wesley Salmon, 'Reply to Pettijohn', *Philosophy of Science* **23** (1956) 150–1.

[666] Wesley Salmon, 'The Predictive Inference', *Philosophy of Science* **24** (1957) 180–90.

[667] Wesley Salmon, 'Inductive Inference', in Baumrin [10] 341–70.

[668] Wesley Salmon, 'Carnap's Inductive Logic', *Journal of Philosophy* **64** (1967) 725–39.

[669] Carl-Erik Sarndal, 'Some Aspects of Carnap's Theory of Inductive Inference', *The British Journal for the Philosophy of Science* **19** (1968) 225–46.

[670] J. T. Saunders, 'Does Knowledge Require Grounds', *Philosophical Studies* **17** (1966) 7–13.

[671] J. T. Saunders and N. Champawat, 'Mr. Clark's Definition of "Knowledge"', *Analysis* **25** (1964) 8–9.

[672] L. J. Savage, 'The Foundations of Statistics Reconsidered', *Proceedings of the Fourth (1960) Berkeley Symposium on Mathematics and Probability*, Berkeley (1961) 575–85.

[673] L. J. Savage, 'Subjective Probability and Statistical Practice', in Savage *et al.* [97] 9–35.

[674] L. J. Savage, 'Implications of Personal Probability for Induction', *Journal of Philosophy* **64** (1967) 593–607.

[675] L. J. Savage, (see Friedman [488]).

[676] I. Scheffler, 'Inductive Inference: a New Approach', *Science* **127** (1958) 177–81.

[677] L. J. Savage, 'A Note on Confirmation', *Philosophical Studies* **11** (1960) 21–3.

[678] L. J. Savage, 'A Rejoinder on Confirmation', *Philosophical Studies* **12** (1961) 19–20.

[679] G. Schlesinger, 'Instantiation and Confirmation', in Cohen and Wartofsky [23].

[680] R. Schwartz, (see Hullett [546]).

[681] Michael Scriven, 'The Principle of Inductive Simplicity', *Philosophical Studies* **6** (1955) 26–30.

[682] Wilfrid Sellars, 'Induction as Vindication', *Philosophy of Science* **31** (1964) 197–231.

[683] Wilfrid Sellars, 'Some Problems about Belief', *Synthese* **19** (1968) 158–77.

[684] Robert K. Shope, 'Explanation in Terms of the Cause', *Journal of Philosophy* **64** (1967) 312–20.

[685] Herbert A. Simon, 'Prediction and Hindsight as Confirmatory Evidence', *Philosophy of Science* **22** (1955) 227–30.

[686] F. Brian Skyrms, 'Nomological Necessity and the Paradoxes of Confirmation', *Philosophy of Science* **33** (1966) 230–49.

[687] F. Brian Skyrms, 'The Explication of "X knows that *p*"', *Journal of Philosophy* **64** (1967) 373–89.

[688] F. Brian Skyrms, (see Rescher [654]).

[689] Robert C. Sleigh, 'A Note on Some Epistemic Principles of Chisholm and Martin', *Journal of Philosophy* **61** (1964) 216–8.

[690] Robert C. Sleigh, 'A Note on Knowledge and Probability', *Journal of Philosophy* **61** (1964) 478.

[691] Robert C. Sleigh, 'A Note on an Argument of Hintikka's', *Philosophical Studies* **18** (1967) 12–4.

[692] Michael A. Slote, 'Some Thought on Goodman's Riddle', *Analysis* **27** (1967) 128–32.

[693] Michael A. Slote, 'A General Solution to Goodman's Riddle?', *Analysis* **29** (1968) 55–8.

[694] Kenneth Small, 'Professor Goodman's Puzzle', *Philosophical Review* **70** (1961) 544–52.

[695] Howard Smokler, 'The Equivalence Condition', *American Philosophical Quarterly* **4** (1967) 300–7.

[696] Howard Smokler, 'Conflicting Concepts of Confirmation', *Journal of Philosophy* **65** (1968) 300–12.

[697] Arthur Smullyan, 'The Concept of Empirical Knowledge', *Philosophical Review* **65** (1956) 362–70.

[698] Joseph D. Sneed, 'Entropy, Information, and Decision', *Synthese* **17** (1967) 329–407.

[699] Vittorio Semezi, 'Can Induction be Mechanized?', *Methodos* **7** (1955) 147–51.

[700] E. Sosa, 'The Analysis of "Knowledge that *p*"', *Analysis* **25** (1964) 1–8.

[701] E. Sosa, 'Propositional Knowledge', *Philosophical Studies* **20** (1969) 33–43.

[702] W. Stegmüller, 'Explanation, Prediction, Scientific Systematization and Nonexplanatory Information', *Ratio* **8** (1966) 1–24.

[703] Alfred J. Stenner, 'A Note on Grue', *Philosophical Studies* **18** (1967) 76–8.

[704] Michael Stocker, 'Knowledge, Causation, and Decision', *Nous* **2** (1968) 65–73.

[705] D. C. Stove, 'Popperian Confirmation and the Paradox of the Ravens', *Australasian Journal of Philosophy* **37** (1959) 149–55.

[706] D. C. Stove, 'A Reply to Mr. Watkins', *Australasian Journal of Philosophy* **38** (1960) 51–4.

[707] D. C. Stove, 'Hempel and Goodman on the Ravens', *Australasian Journal of Philosophy* **43** (1965) 300–10.

[708] D. C. Stove, (see Hooker [545]).

[709] W. A. Suchting, 'Deductive Explanation and Prediction Revisited', *Philosophy of Science* **34** (1967) 41–52.

[710] Patrick Suppes, 'Nelson Goodman on the Concept of Logical Simplicity', *Philosophy of Science* **23** (1956) 153–9.

[711] Patrick Suppes, 'Concept Formation and Bayesian Decisions', in Hintikka and Suppes [42] 21–45.

[712] Patrick Suppes, (see Davidson [475]).

[713] Lars Svenonius, 'Definability and Simplicity', *Journal of Symbolic Logic* **20** (1955) 235–50.

[714] Marshall Swain, 'Skyrms on Non-Derivative Knowledge', *Nous* **3** (1969) 227–31.

[715] R. G. Swineburne, 'Grue', *Analysis* **23** (1968) 123–8.

[716] Judith Jarvis Thomson, 'Grue', *Journal of Philosophy* **63** (1966) 289–309.

[717] Judith Jarvis Thomson, 'More Grue', *Journal of Philosophy* **63** (1966) 528–34.

[718] W. Todd, 'Probability and the Theorem of Confirmation', *Mind* **76** (1967) 260–3.

[719] R. Tuomela, 'Inductive Generalization in an Ordered Universe', in Hintikka and Suppes [42] 155–74.

[720] Joseph Ullian, 'More on "Grue" and Grue', *Philosophical Review* **70** (1961) 386–9.

[721] Peter Unger, 'Experience and Factual Knowledge', *Journal of Philosophy* **64** (1967) 152–73.

[722] Peter Unger, 'An Analysis of Factual Knowledge', *Journal of Philosophy* **65** (1968) 157–70.

[723] John M. Vickers, 'Some Features of Theories of Belief', *Journal of Philosophy* **63** (1966) 197–201.

[724] John M. Vickers, 'Characteristics of Projectible Predicates', *Journal of Philosophy* **64** (1967) 280–6.

[725] R. H. Vincent, 'A Note on Some Quantitative Theories of Confirmation', *Philosophical Studies* **12** (1961) 91–2.

[726] R. H. Vincent, 'The Paradox of Ideal Evidence', *Philosophical Review* **71** (1962) 497–503.

[727] R. H. Vincent, 'Popper on Qualitative Confirmation and Disconfirmation', *Australasian Journal of Philosophy* **40** (1962) 159–66.

[728] John R. Wallace, 'Goodman, Logic, Induction', *Journal of Philosophy* **63** (1966) 310–28.

[729] John R. Wallace, 'Lawlikeness = Truth', *Journal of Philosophy* **63** (1966) 780–1.

[730] G. J. Warnock, 'Claims to Knowledge', *Proceedings of the Aristotelian Society, Supp.* **36** (1962) 19–32.

[731] J. W. N. Watkins, 'Between Analytic and Empirical', *Philosophy* **32** (1957) 112–31.

[732] J. W. N. Watkins, 'A Rejoinder to Professor Hempel's Reply', *Philosophy* **33** (1958) 349–55.

[733] J. W. N. Watkins, 'Mr. Stove's Blunder's', *Australasian Journal of Philosophy* **37** (1959) 240–1.

[734] J. W. N. Watkins, 'A Reply to Mr. Stove's Reply', *Australasian Journal of Philosophy* **38** (1960) 54–8.

[735] J. W. N. Watkins, 'Confirmation without Background Knowledge', *The British Journal for the Philosophy of Science* **10** (1959–60) 318–20.

[736] J. W. N. Watkins, 'Professor Scheffler's Note', *Philosophical Studies* **12** (1961) 16–9.

[737] Jon Wheatley, 'Entrenchment and Engagement', *Analysis* **27** (1967) 119–27.

[738] Frederick J. Will, 'Is There a Problem of Induction?', *Journal of Philosophy* **39** (1942) 505–13.

[739] Frederick J. Will, 'Will the Future be Like the Past?', *Mind* **56** (1947) 332–47.

[740] Frederick J. Will, 'The Justification of Theories', *Philosophical Review* **64** (1955) 370–83.

[741] J. O. Wisdom, 'A Reply to Dr. Das's Criticisms', *The British Journal for the Philosophy of Science* **8** (1957–58) 325–8.

[742] Rollin Workman, 'The Logical Status of the Principle of Induction', *Synthese* **13** (1961) 68–74.

[743] G. H. von Wright, 'The Logical Problem of Induction', *Acta Philosophica Fennica* fasc. **3** (Helsinki 1941). (See Corrigendum, p. 226.)

[744] G. H. von Wright, 'Carnap's Theory of Probability', *Philosophical Review* **60** (1951) 362–74.

CORRIGENDUM

Hilpinen [356] and von Wright [743] are monographs and should be listed in Section I. Technical difficulties prohibit their proper relocation.

ADDENDA

[745] W. H. Baumer, 'In Defense of a Principal Theorem', *Synthese* **20** (1969) 121–42.

[746] G. Spencer Brown, *Probability and Scientific Inference*, London 1957.

[747] M. Deutscher, 'Hintikka's Conception of Epistemic Logic', *Australasian Journal of Philosophy* **47** (1969) 205–8.

[748] B. de Finetti, 'Initial Probabilities: A Prerequisite for any Valid Induction', *Synthese* **20** (1969) 2–16.

[749] I. J. Good, 'Discussion of Bruno de Finetti's Paper: "Initial Probabilities: A Prerequisite for any Valid Induction"', *Synthese* **20** (1969) 17–24.

[750] I. Hacking, 'Linguistically Invariant Inductive Logic', *Synthese* **20** (1969) 25–47.

[751] R. Hilpinen, 'On the Information Provided by Observations', in Hintikka and Suppes [43].

[752] K. J. Hintikka, 'Statistics, Induction, and Lawlikeness: Comments on Dr. Vetter's Paper', *Synthese* **20** (1969) 72–83.

[753] H. Kahane, 'Eliminative Confirmation and Paradoxes', *The British Journal for the Philosophy of Science* **20** (1969) 160–2.

[754] H. E. Kyburg, Jr., *Probability and Induction*, New York (forthcoming).

[755] I. Levi, 'Confirmation, Linguistic Invariance, and Conceptual Innovation', *Synthese* **20** (1969) 48–55.

[756] I. Levi, 'If Jones Only Knew More', *The British Journal for the Philosophy of Science* **20** (1969) 153–9.

[757] I. Levi, Review of Lakatos [58], *Synthese* **20** (1969) 143–8.

[758] C. Pailthrope, 'Knowledge as Justified, True Belief', *Review of Metaphysics* **23** (1969) 25–47.

[759] H. Vetter, 'Logical Probability, Mathematical Statistics, and the Problem of Induction', *Synthese* **20** (1969) 55–71.

Lafayette College

INDEX OF NAMES

INDEX OF SUBJECTS

SYNTHESE LIBRARY

Monographs on Epistemology, Logic, Methodology,
Philosophy of Science, Sociology of Science and of Knowledge, and on the
Mathematical Methods of Social and Behavioral Sciences

‡ROBERT S. COHEN and RAYMOND J. SEEGER, *Ernst Mach: Physicist and Philosopher, Boston Studies in the philosophy of science*. Volume VI. 1970, VIII + 285 pp. Dfl. 38.—

NICHOLAS RESCHER *et al.*, *Essays in Honor of Carl G. Hempel. A Tribute on the Occasion of his Sixty-Fifth Birthday*. 1969, VII + 272 pp. Dfl. 46.—

‡PATRICK SUPPES, *Studies in the Methodology and Foundations of Science. Selected Papers from 1951 to 1969*. 1969, XII + 473 pp. Dfl. 72.—

‡JAAKKO HINTIKKA, *Models for Modalities. Selected Essays*. 1969, IX + 220 pp. Dfl. 34.—

‡D. DAVIDSON and J. HINTIKKA: (eds.), *Words and Objections: Essays on the Work of W. V. Quine*. 1969, VIII+366 pp. Dfl. 48.—

‡J. W. DAVIS, D. J. HOCKNEY, and W. K. WILSON (eds.), *Philosophical Logic*. 1969, VIII + 277 pp. Dfl. 45.—

‡ROBERT S. COHEN and MARX W. WARTOFSKY (eds.), *Boston Studies in the Philosophy of Science*. Volume V: *Proceedings of the Boston Colloquium for the Philosophy of Science 1966/1968*. 1969, VIII + 482 pp. Dfl. 58.—

‡ROBERT S. COHEN and MARX W. WARTOFSKY (eds.), *Boston Studies in the Philosophy of Science*. Volume IV: *Proceedings of the Boston Colloquium for the Philosophy of Science 1966/1968*. 1969, VIII + 537 pp. Dfl. 69.—

‡NICHOLAS RESCHER, *Topics in Philosophical Logic*. 1968, XIV + 347 pp. Dfl 62.—

‡GÜNTHER PATZIG, *Aristotle's Theory of the Syllogism. A Logical-Philological Study of Book A of the Prior Analytics*. 1968, XVII + 215 pp. Dfl. 45.—

‡C. D. BROAD, *Induction, Probability, and Causation. Selected Papers*. 1968, XI + 296 pp. Dfl. 48.—

p.t.o.

‡ROBERT S. COHEN and MARX W. WARTOFSKY (eds.), *Boston Studies in the Philosophy of Science*. Volume III: *Proceedings of the Boston Colloquium for the Philosophy of Science 1964/1966*. 1967, XLIX + 489 pp. Dfl. 65.—

‡GUIDO KÜNG, *Ontology and the Logistic Analysis of Language. An Enquiry into the Contemporary Views on Universals*. 1967, XI + 210 pp. Dfl. 34.—

*EVERT W. BETH and JEAN PIAGET, *Mathematical Epistemology and Psychology*. 1966, XXII + 326 pp. Dfl. 54.—

*EVERT W. BETH, *Mathematical Thought. An Introduction to the Philosophy of Mathematics*. 1965, XII + 208 pp. Dfl. 30.—

‡PAUL LORENZEN, *Formal Logic*. 1965, VIII + 123 pp. Dfl. 18.75

‡GEORGES GURVITCH, *The Spectrum of Social Time*. 1964, XXVI + 152 pp. Dfl. 20.—

‡A. A. ZINOV'EV, *Philosophical Problems of Many-Valued Logic*. 1963, XIV + 155 pp. Dfl. 23.—

‡MARX W. WARTOFSKY (ed.), *Boston Studies in the Philosophy of Science*. Volume I: *Proceedings of the Boston Colloquium for the Philosophy of Science, 1961–1962*. 1963, VIII + 212 pp. Dfl. 22.50

‡B. H. KAZEMIER and D. VUYSJE (eds.), *Logic and Language. Studies dedicated to Professor Rudolf Carnap on the Occasion of his Seventieth Birthday*. 1962, VI + 246 pp. Dfl. 24.50

*EVERT W. BETH, *Formal Methods. An Introduction to Symbolic Logic and to the Study of Effective Operations in Arithmetic and Logic*. 1962, XIV + 170 pp. Dfl. 23.50

*HANS FREUDENTHAL (ed.), *The Concept and the Role of the Model in Mathematics and Natural and Social Sciences. Proceedings of a Colloquium held at Utrecht, The Netherlands, January 1960*. 1961, VI + 194 pp. Dfl. 21.—

‡P. L. R. GUIRAUD, *Problèmes et méthodes de la statistique linguistique*. 1960, VI + 146 pp. Dfl. 15.75

* J. M. BOCHEŃSKI, *A Precis of Mathematical Logic*. 1959, X + 100 pp. Dfl. 15.75

Sole Distributors in the U.S.A. and Canada:

*GORDON & BREACH, INC., 150 Fifth Avenue, New York, N.Y. 10011
‡HUMANITIES PRESS, INC., 303 Park Avenue South, New York, N.Y. 10010